Elements

CARBON

C

Atlantic Europe Publishing

How to use this book

This book has been carefully developed to help you understand the chemistry of the elements. In it you will find a systematic and comprehensive coverage of the basic qualities of each element. Each two-page entry contains information at various levels of technical content and language, along with definitions of useful technical terms, as shown in the thumbnail diagram on the right. There is a comprehensive glossary of technical terms at the back of the book, along with an extensive index, key facts, an explanation of the Periodic Table, and a description of how to interpret chemical equations.

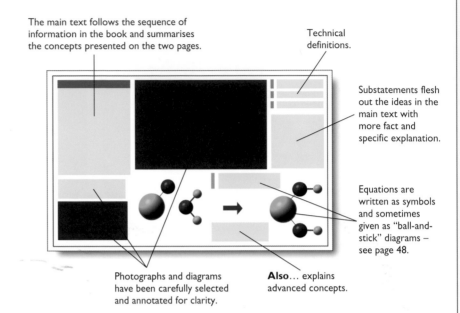

The main text follows the sequence of information in the book and summarises the concepts presented on the two pages.

Technical definitions.

Substatements flesh out the ideas in the main text with more fact and specific explanation.

Equations are written as symbols and sometimes given as "ball-and-stick" diagrams – see page 48.

Photographs and diagrams have been carefully selected and annotated for clarity.

Also… explains advanced concepts.

An Atlantic Europe Publishing Book

Author
Brian Knapp, BSc, PhD
Project consultant
Keith B. Walshaw, MA, BSc, DPhil
        (Head of Chemistry, Leighton Park School)
Industrial consultant
Jack Brettle, BSc, PhD (Chief Research Scientist, Pilkington plc)
Art Director
Duncan McCrae, BSc
Editor
Elizabeth Walker, BA
Special photography
Ian Gledhill
Illustrations
David Woodroffe
Designed and produced by
EARTHSCAPE EDITIONS
Print consultants
Landmark Production Consultants Ltd
Reproduced by
Leo Reprographics
Printed and bound by
Paramount Printing Company Ltd

Suggested cataloguing location
Knapp, Brian
        Carbon
        ISBN 1 869860 49 7
        – Elements series
540

Acknowledgements
The publishers would like to thank the following for their kind help and advice: ICI (UK), Molly and Paul Stratton, Audrey and Teoh Kah Tin, Gutherie Plantation & Agricultural Sdn Bhd and the Kumpulan Gutherie Estate, Catherine and Ian Gledhill, and David Newell.

Picture credits
All photographs are from the **Earthscape Editions** photolibrary except the following:
(c=centre t=top b=bottom l=left r=right)
Courtesy of **ICI(UK)** 26bl and **ZEFA** 41t, 43b.

Front cover: Plastic chips are heated and extruded to make plastic products.
Title page: Adhesives are frequently polymers built from cracked petroleum products.

First published in 1996 by
Atlantic Europe Publishing Company Limited, Greys Court Farm,
Greys Court, Henley-on-Thames, Oxon, RG9 4PG, UK.

This product is manufactured from sustainable managed forests. For every tree cut down at least one more is planted.

The demonstrations described or illustrated in this book are not for replication. The Publisher cannot accept any responsibility for any accidents or injuries that may result from conducting the experiments described or illustrated in this book.

Contents

Introduction

An element is a substance that cannot be broken down into a simpler substance by any known means. Each of the 92 naturally occurring elements is therefore one of the fundamental materials from which everything in the Universe is made. This book is about the element carbon.

Carbon

Carbon, the sixth most common element on Earth, is an essential part of nearly all living things. About 94% of the six million known compounds contain carbon, far more than any other element.

Carbon, in the form of diamond, is the hardest natural material on Earth. Carbon is found in a wide variety of rocks, such as chalk and limestone. Compounds of carbon such as coal, natural gas and oil, provide the world's most important fuels.

Carbon dioxide gas is one of the main compounds in the air, playing a vital role in controlling the temperature of the atmosphere (it is what we call a "greenhouse gas"). Carbon dioxide is also found dissolved in all water.

Carbon is an essential element in many of the man–made, or synthetic, substances that we use in the modern world, for example, plastics, all synthetic fabrics, many medicines and a wide variety of other chemicals.

Not surprisingly, perhaps, among this huge variety of substances, most of which are good for

our way of living, there will be some that cause concern. Carbon-containing compounds called CFCs have been responsible for destroying part of the world's protective ozone layer. Other compounds, such as the pesticide DDT, may be life-threatening if humans and animals are exposed to large doses.

Historically, as scientists tried to classify the bewildering array of carbon-based compounds, they made an important distinction between those they called organic compounds, and those called inorganic compounds. Limestone rock (calcium carbonate) and carbon gases (such as carbon dioxide and carbon monoxide) are part of the group called inorganic chemicals. Chemicals like those based on petroleum, which are formed from living tissue, are by far the most numerous and are called organic chemicals.

Because carbon compounds so greatly outnumber those made of any other element, it would be impossible to include examples of the majority of the compounds and their properties. Rather, this book will describe the main categories of carbon compounds and a few of the most common compounds.

◄ This is silicon carbide, a compound of silicon and carbon, SiC, whose crystals form black, glossy plates. It is an extremely hard mineral and known by the common name of carborundum. When ground down into a grit, carborundum makes an abrasive that is widely used in industry.

What do carbon compounds have in common?

Carbon is part of many compounds, all of which have certain common properties. For example:

❶ Few carbon based-compounds change quickly at ordinary temperatures, but they begin to react fiercely at high temperatures (as in burning).

❷ All carbon compounds that form tissues – plants, tar, oil, natural gas, etc. – will burn (they are combustible) and can be used as a fuel. When they burn, the compound is oxidised and carbon dioxide is produced. The remaining material is nearly pure carbon (which can be seen in the charred nature of burnt wood, for example).

❸ Many carbon-based compounds are not attracted to water and so in general do not dissolve in water. As a result, water alone cannot be used to remove grease or oil from a surface, nor will water dissolve our skin, because all of these things are carbon compounds.

❹ Groups that contain carbon and nitrogen often have an unpleasant smell in liquid form. Some people compare it to rotting fish. Such smells are mainly confined to the factories where the materials are made. The common fabric material nylon, for example, which is a plastic that contains nitrogen, has no smell once it is made into a yarn.

❺ Some compounds of carbon and nitrogen are very unstable and can be made into explosives. Two of the more common explosives are TNT (trinitrotoluene) and nitroglycerine (glycerol trinitrate).

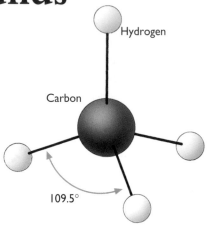

Hydrogen

Carbon

109.5°

▲ This is a model of a hydrocarbon molecule. It consists of a central carbon atom linked (bonded) to four hydrogen atoms to form the molecule methane (CH_4).

What is so special about carbon?

Carbon is very special because it can form so many compounds. The explanation lies deep inside the atom. Carbon atoms can form strong links with four other atoms. This dramatically increases the number of patterns that carbon atoms can make.

Carbon can also link together in long chains or rings, carbon to carbon to carbon to carbon and so on. Chemists call these links chemical bonds; very long chains, made by joining short ones, are called polymers. And, quite unusually, these long chains cannot be destroyed by water or air or be attacked by bacteria. This explains why so many plastics do not disintegrate in the environment in the way that other materials do. Only sunlight can destroy some polymers, causing the chains to break, and the material to become brittle.

combustion: the special case of oxidisation of a substance where a considerable amount of heat and usually light are given out. Combustion is often referred to as "burning".

compound: a chemical consisting of two or more elements chemically bonded together. Calcium atoms can combine with carbon atoms and oxygen atoms to make calcium carbonate, a compound of all three atoms.

plastic (material): a carbon-based material consisting of long chains (polymers) of simple molecules. The word plastic is commonly restricted to synthetic polymers.

plastic (property): a material is plastic if it can be made to change shape easily. Plastic materials will remain in the new shape. (Compare with elastic, a property where a material goes back to its original shape.)

▶ Butane gas is a hydrocarbon and contains carbon and hydrogen bonded together. When set alight it burns, producing heat.

◀ Living organisms depend on carbon atoms for their existence. Tissues and bones or shells and wings are all carbon-based compounds.

◀ This is a piece of anthracite, a form of hard coal with a very high carbon content. Anthracite was formed by the slow decomposition of plant tissue under high temperature and pressure, deep within the Earth.

The Carboniferous Period is a part of geological time that began about 400 million years ago and lasted for some 60 millions years. In the 19th century geologists working in Britain recognised that much of the coal they were mining was formed during the same part of the Earth's history. Because coal is a carbon-based compound, they called the entire period the Carboniferous.

In fact, the Carboniferous Period contains carbon-based rocks of two kinds. During the early part, many thick beds of limestone were laid down. These Carboniferous limestones are made of calcium carbonate and so are carbon-based rocks. They also contain large amounts of oil and natural gas (petroleum).

During the later part of the Carboniferous Period, the main coal-bearing rocks on Earth were formed, together with more oil and natural gas (petroleum).

Geologists now know that coal and petroleum deposits have formed in a great many geological periods, and are still forming today.

Crystals of carbon

Three minerals – graphite, diamond and the more recently discovered buckminsterfullerene (known as "buckyballs") – are made solely of carbon atoms. Of these, graphite is the most common. It occurs in rocks, and it is also formed as small crystals when hydrocarbons burn in the absence of air (e.g. coke, charcoal).

Diamond is far more rare than graphite. Diamonds were formed under immense temperatures and pressures, such as found in pipes leading to ancient volcanoes. The most famous diamond mine, at Kimberley, South Africa, follows an old volcanic pipe for more than two kilometres vertically into the Earth.

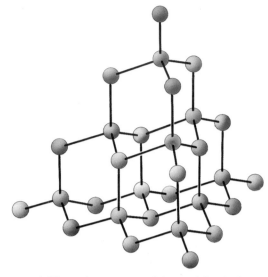

▲ This is the structure of diamond. It is built of interlocking carbon atoms with no room for other atoms to form part of the structure. This is what makes the mineral so unreactive.

Diamond: sparkling crystals of carbon

Carbon atoms can link to form a very stable mineral. Diamonds have atoms so tightly bonded together that they are one of the hardest substances known.

Pure diamond is colourless and transparent. It commonly forms a shape like two pyramids base to base (a tetrahedron). Jewellers make use of this property when they cleave rough diamonds to make jewellery. Each of the faces (called facets) is created by splitting the diamond parallel to the faces of its crystals.

Diamond is not always colourless, and if it contains impurities it may be a darker colour. Some diamonds are almost black.

◀ This piece of Kimberlite rock shows the way that most diamond occurs, as a dull yellowish mineral set in a rock background. This diamond is translucent and highly flawed. Only occasionally does a transparent and flawless piece of mineral occur. The cut diamond placed on the rock shows the comparison.

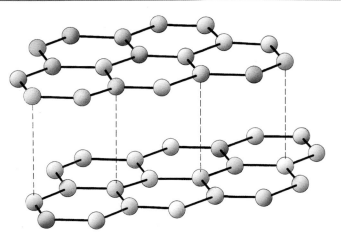

crystal: a substance that has grown freely so that it can develop external faces. Compare with crystalline, where the atoms are not free to form individual crystals and amorphous where the atoms are arranged irregularly.

electrode: a conductor that forms one terminal of a cell.

mineral: a solid substance made of just one element or chemical compound. Calcite is a mineral because it consists only of calcium carbonate, halite is a mineral because it contains only sodium chloride, quartz is a mineral because it consists of only silicon dioxide.

▲▶ This is the structure of graphite. It is made only of carbon minerals and is, like diamond, an unreactive substance. However, because the structure is in sheets, the bonds between the sheets are relatively weak, so that when pressure is applied, parts of the mineral flake off. This is what allows graphite to be used in pencils.

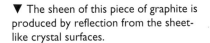

▼ The sheen of this piece of graphite is produced by reflection from the sheet-like crystal surfaces.

Graphite

Graphite is a black, soft form of carbon, harder than coal but far softer than diamond. Graphite naturally crumbles to release tiny flakes. This occurs because in graphite the carbon atoms are arranged in sheets that are poorly linked to each other. As a result, one sheet of crystals readily slides over another. It is this property that makes graphite useful as a lubricant (it is used to lubricate door locks, for example) and also as a pencil "lead" because it leaves a trail of black flakes as it is moved across paper.

Graphite can also conduct electricity and heat. Graphite makes the central electrode in dry batteries in many pieces of electrical equipment, such as the brushes in electric motors where sparks occur, as well as in huge steel furnaces.

The carbon cycle

The carbon cycle is the name given to the way that carbon is transferred between plants, animals, the atmosphere, rocks and oceans.

The carbon cycle is crucial to the way the planet works. It is a very complicated cycle, of which a simplified form is shown here.

The main reservoirs of carbon are in the air (in the form of carbon dioxide) and in rocks (either as limestone and chalk or as oil, natural gas and coal). The main transfers occur through the growth and death of plants (involving the chemical processes of photosynthesis, respiration and oxidation) and through the way people burn fossil fuels (involving the process of oxidation).

Animals

The proteins created by plants are used to make the tissues of animals, and they also provide sources of sugars, starches and fats. To release energy locked up as glucose, the sugars, starches and fats are combined with oxygen (they are oxidised). As sugars are oxidised, carbon dioxide and water are returned to the environment as part of the carbon cycle. As cells wear out or die, they are broken down into carbon dioxide, water and other simple compounds.

For the most part, plant and animal tissues are broken down by decomposing organisms and the elements returned to the environment for reuse. However, in some cases, tissues are not immediately broken down and recycled, but get buried and are preserved. This produces a store of energy in the form of fossil fuels. In this case the carbon cycle is interrupted until the fuels are burned or until erosion exposes them at the surface, at which point they oxidise.

Respiration

All animals produce carbon dioxide in their lungs. The lungs breathe in air containing oxygen and expel air containing carbon dioxide. This is called respiration.

Plants release carbon dioxide gas as they convert sugars for energy. This is also respiration.

> **EQUATION: Oxidation of glucose**
>
> *Glucose + oxygen ⇨ carbon dioxide + water*
>
> $C_6H_{12}O_6(s) + 6O_2(g) \Rightarrow 6CO_2(g) + 6H_2O(l)$

▼ This diagram represents the way that sugars are oxidised during respiration to produce carbon dioxide gas. During photosynthesis this reaction is reversed.

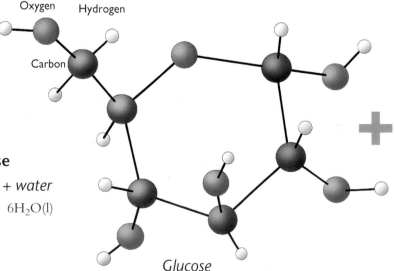

Oxygen   Hydrogen

Carbon

Glucose

Plants

Only living plants can make compounds of carbon. They use the energy of sunlight in a process called photosynthesis.

In photosynthesis, six molecules of carbon dioxide from the air combine with six molecules of water, forming one molecule of glucose (sugar), and releasing six molecules of oxygen back into the atmosphere.

Some sugar is combined with nitrogen compounds to form the proteins that make up tissues.

Sugar is also converted into larger, more complex molecules called starches and fats. These are forms of energy that can be stored for later use.

fossil fuels: hydrocarbon compounds that have been formed from buried plant and animal remains. High pressures and temperatures lasting over millions of years are required. The fossil fuels are coal, oil and natural gas.

molecule: a group of two or more atoms held together by chemical bonds.

oxidation: a reaction in which the oxidising agent removes electrons.

photosynthesis: the process by which plants use the energy of the Sun to make the compounds they need for life. In photosynthesis, six molecules of carbon dioxide from the air combine with six molecules of water, forming one molecule of glucose (sugar) and releasing six molecules of oxygen back into the atmosphere.

Plants make sugars by photosynthesising sugars using carbon dioxide from the air.

As sugars are oxidised by plants for energy, carbon dioxide is released back into the air.

The main source of carbon dioxide is the air.

As sugars are oxidised by animals for energy, carbon dioxide is released back to the air.

Plants release carbon dioxide when they are burned (a rapid form of decay).

Carbon-based chemicals are absorbed by animals as they consume plant matter.

► The natural carbon cycle. Note that, in addition, over geologic time carbon is stored in rocks. It is released when fossil fuels are burned.

Organisms in the soil break down dead plant and animal tissue, and carbon dioxide gas is released as it decays.

EQUATION: Photosynthesis

Carbon dioxide + water ⇨ glucose + oxygen

$$6CO_2(g) + 6H_2O(l) \Rightarrow C_6H_{12}O_6(s) + 6O_2(g)$$

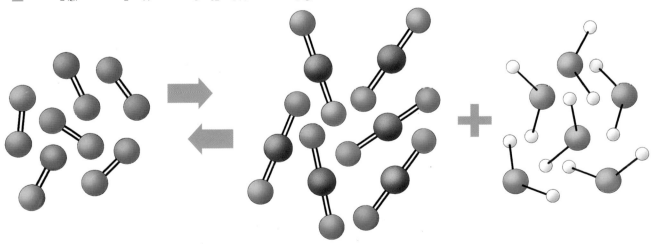

Oxygen   Carbon dioxide   Water

Carbon dioxide in the environment

Carbon dioxide gas (like water vapour) strongly absorbs, or soaks up, the heat that is radiated by the Earth out into space. Without carbon dioxide and water vapour, the Earth would be a very cold, almost uninhabitable place, some 25°C colder than it is today.

The amount of heat stored changes with the amount of carbon dioxide gas in the air. Over the last few centuries, people have increasingly burned more fuel, so much more carbon dioxide has been added to the air than was previously the case.

The extra carbon dioxide has been able to absorb even more of the earth's heat, leading to a gradual warming of the air. This process, known as the Greenhouse Effect, has been charted for more than a century.

Some scientists are concerned about the build up of carbon dioxide because they fear that "global warming" will cause all sorts of unpredictable changes, such as droughts and floods, as well as a possibly disastrous rise in the sea level as the warm air causes the Antarctic ice sheet to melt.

▲ Plants need carbon to make the cells of their bodies. They extract carbon from carbon dioxide gas using the energy in sunlight.

Also...

The Earth absorbs energy through solar radiation. This is mostly short wave radiation. The gases in the atmosphere are largely transparent to solar radiation. The incoming solar radiation does little to warm the air, therefore, and mainly warms the land and oceans.

The wavelength of radiation depends on the temperature of the radiating body. The Sun is very hot and so radiates in short wavelengths. However, the Earth is cool and radiates in longer (infra-red) wavelengths.

Water vapour and carbon dioxide gas both absorb in the infra-red wavelengths better than in short wavelengths, so they absorb some of the heat energy that would otherwise be lost to space. This is what makes the atmosphere warmer and causes the Greenhouse Effect.

► The passage of long wave radiation from the earth and back to space is slowed by the increasing "blanket" of carbon dioxide in the atmosphere as more of the gas is released by burning fossil fuels.

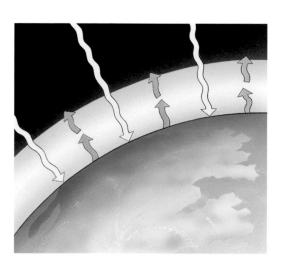

Greenhouse Effect: an increase of the global air temperature as a result of heat released from burning fossil fuels being absorbed by carbon dioxide in the atmosphere.

vapour: the gaseous form of a substance that is normally a liquid. For example, water vapour is the gaseous form of liquid water.

► The decay of organic matter is important in releasing carbon to the cycle. Normally leaves decay within a few years, which is why there is rarely thick leaf litter on the ground. Only where the ground is permanently waterlogged, such as in still-water marshes and bogs, can organic material accumulate. When it does so, it locks up carbon. Major swamps have occurred in many periods of geological history. These eventually became buried, forming coal deposits, oil and natural gas reserves.

▼ Clear-cutting trees has a major effect on the natural carbon cycle because it removes a natural carbon-absorbing part of the environment (called a carbon "sink"). To maintain balance in the carbon cycle, trees have to be replanted in the same quantities that they are felled. This is happening in the temperate lands, but not in the tropics.

Using carbon dioxide

Carbon dioxide is sometimes used in its solid form – as dry ice – to create the impression of fog. It is safe, clean and easy to control in theatres because of three important properties: carbon dioxide sublimes (changes straight from a solid to a gas and vice versa), is heavier than air and is noncombustible.

Carbon dioxide gas is produced during the preparation of both food and drink, in some cases for effect (as in carbonated water), in other cases to "lighten" dough, cakes and pastries.

Baking

The lightness of many baked flour-based foods depends on the dough, pastry, etc., containing a large number of bubbles of carbon dioxide gas. This can be achieved by heating baking soda (sodium bicarbonate), causing it to decompose, and thus releasing carbon dioxide gas.

Baking powder is a mixture of baking soda (sodium bicarbonate), tartaric acid, and small amounts of starch. As soon as the acid and the baking soda are wetted they begin to react, releasing carbon dioxide. The amount and size of the bubbles depend on the rate at which the gas bubbles are created, larger bubbles being released by greater amounts of gas.

Carbon dioxide gas is usually used to lighten the texture of (or leaven) doughs, sweet pastries and cakes.

▲ Fizzy drinks contain carbonated water, that is, water in which carbon dioxide has been dissolved under pressure. In some cases the carbon dioxide is formed in the bottle or can by mixing an acid (such as phosphoric acid) and an alkali with the drink just before it is sealed in its container. These react to produce gas, although the gas remains dissolved while the container is closed.

When the cap is loosened, the pressure is reduced and the carbon dioxide comes out of solution, producing bubbles: the fizz.

▶ This cake shows how carbon dioxide bubbles form from the reaction of baking powder. The bubbles become trapped by the stickiness of the dough, thus giving cakes their light texture.

Dry ice

Frozen carbon dioxide is called dry ice and is used for portable refrigeration and to create swirling fog on theatre stages or movie sets.

EQUATION: Producing carbon dioxide by heating baking soda

Sodium bicarbonate ⇨ sodium carbonate + carbon dioxide + water

$$2NaHCO_3(s) \quad \Rightarrow \quad Na_2CO_3(s) \quad + \quad CO_2(g) \quad + \quad H_2O(l)$$

Preparation of carbon dioxide gas

Carbon dioxide gas can be produced in numerous ways in the laboratory. For example, if dilute hydrochloric acid is added to calcium carbonate, it begins to fizz. The reaction produces a solution of calcium chloride and releases carbon dioxide gas.

gelatinous: a term meaning made with water. Because a gelatinous precipitate is mostly water, it is of a similar density to water and will float or lie suspended in the liquid.

noncombustible: a substance that will not burn.

reagent: a starting material for a reaction.

sublimation: the change of a substance from solid to gas, or vice versa, without going through a liquid phase.

EQUATION: Preparation of carbon dioxide gas

Dilute hydrochloric acid + calcium carbonate ⇨ calcium chloride + carbon dioxide + water

$$2HCl(aq) \quad + \quad CaCO_3(s) \quad ⇨ \quad CaCl_2(aq) \quad + \quad CO_2(g) \quad + \quad H_2O(l)$$

▼ Some antacid tablets contain calcium carbonate, which acts on the dilute hydrochloric acid in the stomach. This demonstration shows that a reaction occurs in the stomach, releasing carbon dioxide gas. This explains why some antacid tablets cause people to "burp".

Carbon dioxide as a fire extinguisher gas

Because carbon dioxide gas is noncombustible, it is ideal for use in extinguishing fires.

Fire extinguishers may contain two reagents (liquids that will react) to produce carbon dioxide quickly. The equation below shows the reaction of aluminium sulphate and sodium carbonate. These reagents were used as a source of carbon dioxide gas and a source of foam for many years until they were replaced by pressurised carbon dioxide cylinders.

In use, the two reagents are kept apart inside the extinguisher. When in use, a knob on the extinguisher is struck, breaking the seal between the liquids and causing them to react.

The reaction produces a gelatinous liquid and carbon dioxide gas. This gas cannot easily escape through this sticky liquid, and instead forms bubbles inside it. The result is that a foam containing carbon dioxide immediately issues from the extinguisher nozzle. This has the effect of blanketing the fire with materials that will not burn. It also prevents oxygen from reaching the flames.

Also...

Many modern carbon dioxide extinguishers do not rely on a chemical reaction, but instead contain carbon dioxide gas under pressure in a strong cylinder.

EQUATION: Fire extinguishing

Aluminium sulphate (alum) + sodium carbonate + water ⇨ aluminium hydroxide (gelatinous precipitate) + carbon dioxide (gas) + sodium sulphate (solution)

$$Al_2(SO_4)_3(aq) \quad + \quad 3Na_2CO_3(aq) \quad + \quad 3H_2O(l) \quad ⇨ \quad 2Al(OH)_3(s) \quad + \quad 3CO_2(g) \quad + \quad 3Na_2SO_4(aq)$$

Carbon monoxide

Carbon monoxide is a reducing agent, that is it takes oxygen from some materials it contacts. This is a useful reaction, and carbon monoxide is used widely in industry, especially in the refining of metals from their ores. For example, the chemical reactions inside a blast furnace involve the reduction of iron ore to iron metal.

However, carbon monoxide is also produced when fuels are burned because the burning is not usually efficient enough to produce only carbon dioxide. Unless exhaust gases from burning fuels are allowed to disperse, this colourless, tasteless and odourless gas can build up and be fatal.

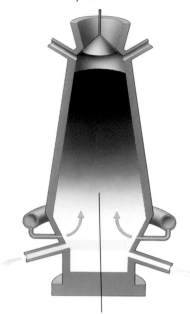

▼ A blast furnace in which iron oxide is reduced by carbon monoxide.

Carbon monoxide acts as a reducing agent in the heart of the furnace.

▼ A diagram of the way carbon monoxide is produced during fuel ignition.

The petrol and air mixture is compressed by the piston and ignited by an electric spark from the spark plug. The reaction is explosive and forces the piston down, completing a stroke.

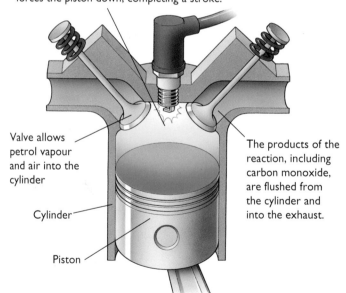

Valve allows petrol vapour and air into the cylinder

The products of the reaction, including carbon monoxide, are flushed from the cylinder and into the exhaust.

Cylinder

Piston

Also...

Catalytic converters used in motor-vehicle exhausts are designed in part to cause the conversion of carbon monoxide to carbon dioxide and water. They achieve this by using the oxygen from another exhaust gas, nitric oxide. Single atoms of oxygen are very reactive and so readily combine with carbon monoxide to form carbon dioxide.

Inefficient combustion

In a typical internal combustion engine, the fuel is a hydrocarbon in the form of a mist of tiny droplets. When mixed with air and ignited with a hot object such as the spark plug in a car engine, the hydrocarbon reacts violently to produce hot gas.

If the engine were completely efficient, all the energy in the hydrocarbon would be turned into power to drive the cylinders. At the same time, the carbon would turn into harmless carbon dioxide. But no engine is very efficient, and the carbon does not burn up completely. As a result, carbon monoxide gas is also produced.

Carbon monoxide can be fatal because, when breathed into the lungs, the red blood cells absorb it instead of oxygen. As a result oxygen cannot get to the brain. For this reason good ventilation is needed whenever engines are working inside a building.

Carbon monoxide in the atmosphere slowly changes to carbon dioxide as it combines with oxygen in the air.

EQUATION: Iron oxide reduced by carbon monoxide

Iron oxide + carbon monoxide ⇨ iron metal + carbon dioxide

$$Fe_2O_3(s) \quad + \quad 3CO(g) \quad ⇨ \quad 2Fe(l) \quad + \quad 3CO_2(g)$$

catalyst: a substance that speeds up a chemical reaction but itself remains unaltered at the end of the reaction.

oxidation/reduction: a reaction in which oxygen is gained/lost.

Carbon monoxide as a reducing agent

This laboratory demonstration shows how carbon monoxide can act as a reducing agent, taking oxygen from a metal ore. It is an example of a refining technique widely used in industry.

The glass tube contains black copper oxide. Carbon monoxide gas is blown through the tube, and the surplus is ignited and burns with a blue flame from a small hole in the tube.

Notice how the copper changes to an orange colour. The heat of the Bunsen burner is speeding up the reaction of the carbon monoxide with the copper oxide, producing carbon dioxide gas and leaving pure copper behind.

The way that carbon monoxide takes up oxygen can also be used in industry to refine metal. A blast furnace, used to produce iron from iron ore, uses coke and air to produce the carbon monoxide and the heat needed to release the iron from its ore.

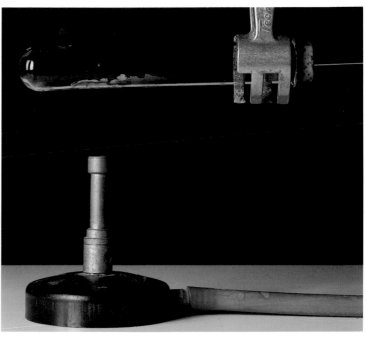

❶▲ Carbon monoxide is passed over the black copper oxide powder (copper ore) in a test tube. The glowing surface shows where the reaction between copper oxide and carbon monoxide is occurring. As the carbon monoxide exits the tube it is burnt giving a characteristic blue flame.

❷▼ The reduced copper oxide changes colour as the oxygen is removed. The material left is pure copper.

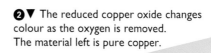

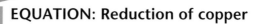

EQUATION: Reduction of copper

Copper oxide + carbon monoxide ⇨ carbon dioxide + copper

$$CuO(s) \quad + \quad CO(g) \quad \underset{heat}{⇨} \quad CO_2(g) \quad + \quad Cu(s)$$

Carbon in food

Carbohydrate is the name for a wide range of natural compounds such as sugar and starch, containing carbon, hydrogen and oxygen.

The simplest carbohydrate is glucose. This in turn is used to make proteins that build plant tissues called cellulose. Soft forms of cellulose make up the fleshy parts of leaves, for example, while a harder, reinforced form of cellulose called lignin, makes up the veins of leaves, twigs, bark, etc.

Carbohydrates are a store of energy used both by plants and by animals when they eat plants. In this way the energy from the Sun is converted into energy for all living things.

Because they are so rich in energy, carbohydrates form the main part of most people's diets.

Fossilised carbohydrates (fossil plant tissues) are also the source of the fuels burned as coal and oil.

❶▲ Concentrated sulphuric acid is added to white sugar.

❷▶ It begins to froth and turn brown. This is an exothermic reaction, so a considerable amount of heat is given off in the water, producing steam.

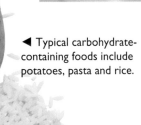

◀ Typical carbohydrate-containing foods include potatoes, pasta and rice.

Oxidation of carbohydrates

The body makes use of carbohydrates by the process of oxidation. It is the reverse of photosynthesis. Plants store the carbohydrate glucose as starch and sucrose. In animals, glucose is sent around the body in the bloodstream, and any excess is converted to fat and stored for later use.

EQUATION: Oxidation of glucose

Glucose + oxygen ⇨ water + carbon dioxide (+ a release of energy)

$C_6H_{12}O_6(s)$ + $6O_2(g)$ ⇨ $6H_2O(l)$ + $6CO_2(g)$ (+ energy released)

Sulphuric acid dehydrates sugars

If sulphuric acid is added to sugar (sucrose) the sugar dehydrates, that is, it loses its water and turns into black carbon.

The reaction produces considerable heat, so water is released as steam. The chemical equation shows that the sulphuric acid remains uncombined.

As the steam is given off, bubbles form, which cause the carbon to develop into a "volcano" of a substance which, on cooling, has the feel of coke.

dehydration: the removal of water from a substance by heating it, placing it in a dry atmosphere, or through the action of a drying agent.

glucose: the most common of the natural sugars. It occurs as the polymer known as cellulose, the fibre in plants. Starch is also a form of glucose. The breakdown of glucose provides the energy that animals need for life.

oxidation: a reaction in which the oxidising agent removes electrons. (Note that oxidising agents do not have to contain oxygen.)

❸▼▶ As the glucose is dehydrated it changes to black carbon. At this stage the glucose is a hot, syrupy liquid, which does not readily allow the steam to escape, so some of it remains trapped as bubbles.

❹▼ The expanding bubbles are trapped in the carbon as the reaction finishes and the carbon cools to a coke-like hard material.

EQUATION: Dehydration of sucrose

Sucrose + sulphuric acid ⇨ water + carbon + sulphuric acid

$C_{12}H_{22}O_{11}(s) + H_2SO_4(aq) ⇨ 11H_2O(g) + 12C(s) + H_2SO_4(aq)$

Separating carbon compounds

Organic materials, those containing carbon, are mainly very chemically complex; however, this is not always noticable. For example, the red juice from, say, a beetroot, may not at first sight seem complex at all.

But, just as there are ways of showing that the simple colour of white light is in fact made of many colours of light combined, so it is possible to show that a "simple" juice is made of an extraordinary array of chemicals, each one containing different carbon compounds. On these pages you can see how it is done using the process of chromatography.

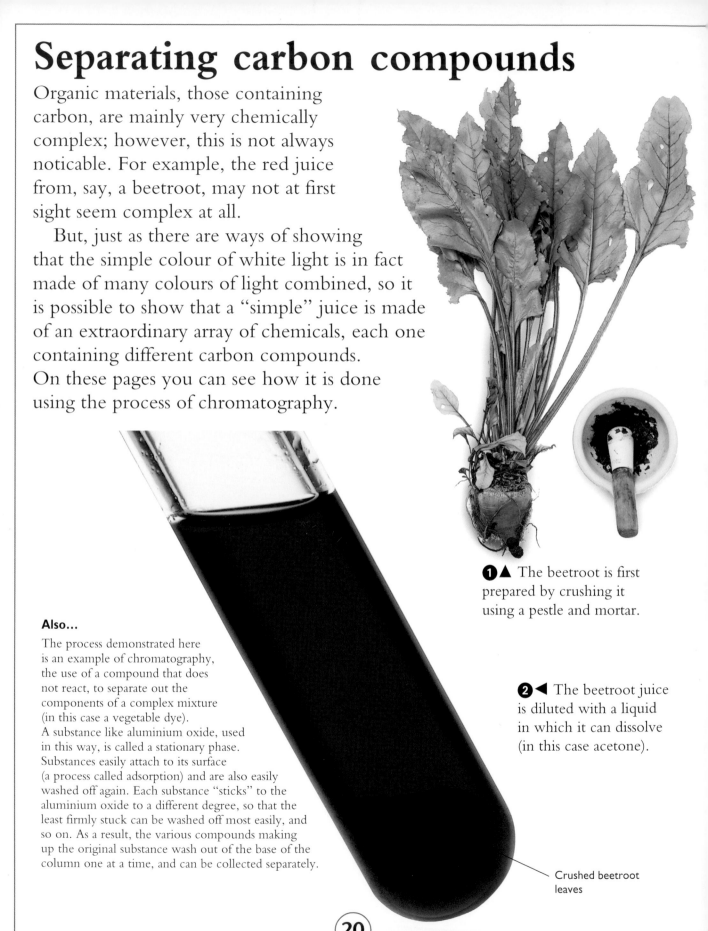

❶▲ The beetroot is first prepared by crushing it using a pestle and mortar.

❷◀ The beetroot juice is diluted with a liquid in which it can dissolve (in this case acetone).

Also...

The process demonstrated here is an example of chromatography, the use of a compound that does not react, to separate out the components of a complex mixture (in this case a vegetable dye). A substance like aluminium oxide, used in this way, is called a stationary phase. Substances easily attach to its surface (a process called adsorption) and are also easily washed off again. Each substance "sticks" to the aluminium oxide to a different degree, so that the least firmly stuck can be washed off most easily, and so on. As a result, the various compounds making up the original substance wash out of the base of the column one at a time, and can be collected separately.

Crushed beetroot leaves

Acetone

acetone: a petroleum-based solvent.

mixture: a material that can be separated out into two or more substances using physical means.

3 ◀ A column of aluminium oxide is prepared by filling it with acetone. The aluminium oxide acts as a chemical filter, releasing the components of the juice one at a time.

Aluminium oxide

Cotton wool support

Rubber stopper with hole

4 ◀ The first component of the beetroot juice emerges from the base of the tube. It is a yellow substance.

5 ▼ These are the first two components collected from the tube. Notice that they are slightly different colours. The substance on the right is chlorophyll (the green pigment in plants), while that on the left is called xanthophyll.

Separated components of the beetroot drip from the the column into a test-tube.

Charcoal

Charcoal is wood that has been burned at about 1000°C in the absence of air. It is almost pure carbon and consists of tiny crystals of graphite.

Charcoal is able to burn at much higher temperatures than wood, and it is smokeless, so it makes a good fuel.

Activated charcoal is a form of charcoal made by burning waste organic matter (twigs, wood bark, sawdust, etc.) in the absence of air. When it has been processed it has an enormous surface area that is able to absorb molecules of gas. For this reason it is often used in situations where gas molecules need to be absorbed.

Carbon black

When people are asked what colour they associate with carbon, they usually say black. If people are asked to name a common form of carbon, they often say soot.

In fact black soot (called lampblack) from badly adjusted oil lamps, for example, is almost pure carbon. Pure carbon, a deep black powder, such as is used in photocopiers and laser printers, is obtained by heating anything containing carbon – coal and wood, for example – in a furnace where there is no air. As a result, coal changes to coke and wood to charcoal.

Powdered carbon contains spheres of the rare form of carbon called buckminsterfullerene. Powdered carbon is used in rubber and plastics to slow down the rate at which these materials deteriorate in sunlight. Carbon powder is also used to make black ink and paint.

▲ Charcoal glows when it is heated with a gas jet from a Bunsen burner. Notice that the charcoal burns without any smoke; all of the graphite crystals are fused together in the charcoal and thus are not released as soot particles.

▼ Because it is made from wood, charcoal is an attractive fuel in the developing world where people need a fuel that provides a high temperature for cooking but where they cannot afford to use electricity, bottled gas or paraffin. Charcoal is also made into briquettes and used for barbecue fires in industrial countries.

❶ ◄ A piece of activated charcoal has been dropped into a gas jar containing bromine and the cover glass replaced.

❷ ▼ Within a minute the colour of the gas is getting lighter as fewer free bromine molecules remain in the jar.

adsorb: to "collect" gas molecules or other particles on to the *surface* of a substance. They are not chemically combined and can be removed. (The process is called "adsorption".) Compare to absorb.

halogen: one of a group of elements including chlorine, bromine, iodine and fluorine.

❸ ▼ After a few minutes the gas jar is colourless, because all of the bromine molecules are now adsorbed on to the surface of the activated charcoal with none remaining as free gas.

Activated charcoal

Activated charcoal has a very large reactive surface area (about 2000 sq m of surface area for every gram in weight of charcoal). It is able to soak up (adsorb) large numbers of gas molecules on this vast surface.

This impressive property means that activated charcoal has been widely used as a gas filter, from gas masks for use in war or in fire-fighting, to removing unpleasant odours. It is also used in water purification plants.

This sequence of pictures shows a gas jar, with activated charcoal in the bottom, that has been filled with bromine, a poisonous, brown halogen gas. The pictures were taken over a few minutes. Notice how the amount of free bromine in the gas jar (as seen by the colour of the gas) decreases. In the gas jar on the right, there is no free bromine left at all.

Once all the sites on the activated charcoal have been used, it has to be thrown away. It cannot be reactivated. Although activated charcoal works well for many hydrocarbons, chlorine and similar gases, carbon will not absorb oxygen or nitrogen. Gas-suits with activated charcoal linings were used by the allied forces in the Gulf War because there was a threat from gas attack.

Hydrocarbons

The word petroleum comes from the Latin words *petra* and *oleum*, meaning "rock" and "oil", respectively. Petroleum is a "catch-all" name for a range of hydrocarbon gases, liquids and some solids, which form in the rocks of the Earth's crust.

Petroleum is usually a complicated mixture of liquids, solids and dissolved gases. The liquid form of petroleum is referred to as crude oil. The gases associated with petroleum are called natural gas. From these liquids and gases come the fuels that power the modern world and the raw materials for plastics, fertilisers, drugs and a wide range of other essential materials.

▲ Natural gas is predominantly made of the hydrocarbon methane, CH_4

▼ Crude oil is a brown, often sulphurous-smelling liquid. This sample is from the second well ever drilled in Texas.

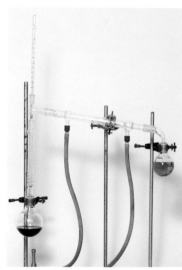

▲ The laboratory apparatus shown in the diagram on the right looks like this.

Also...

The most common carbon-based compounds contain simply hydrogen and carbon, derived from living tissue. These are called hydrocarbons and they are the basis of most fuels such as coal, oil and gas.

Other groups of carbon-based compounds include alcohols such as ethanol, the substance found in all alcoholic beverages; acetone, which can dissolve many plastics; and esters, which smell of fruits.

Distilling hydrocarbons in the laboratory

Crude oil is a mixture of a variety of components, each with a different boiling point. The easiest way to separate out the components is to heat the oil.

The principle of the process is shown in this laboratory glassware setup. The flask of oil is heated, which, depending on the degree of heating, will vaporize a range of compounds.

As the vapours rise through the glassware, the "lightest" fraction escapes from the top of the fractionating column and enters the condenser.

crude oil: a chemical mixture of petroleum liquids. Crude oil forms the raw material for an oil refinery.

fraction: a group of similar components of a mixture. In the petroleum industry the light fractions of crude oil are those with the smallest molecules, while the medium and heavy fractions have larger molecules.

petroleum: a natural mixture of a range of gases, liquids and solids derived from the decomposed remains of plants and animals.

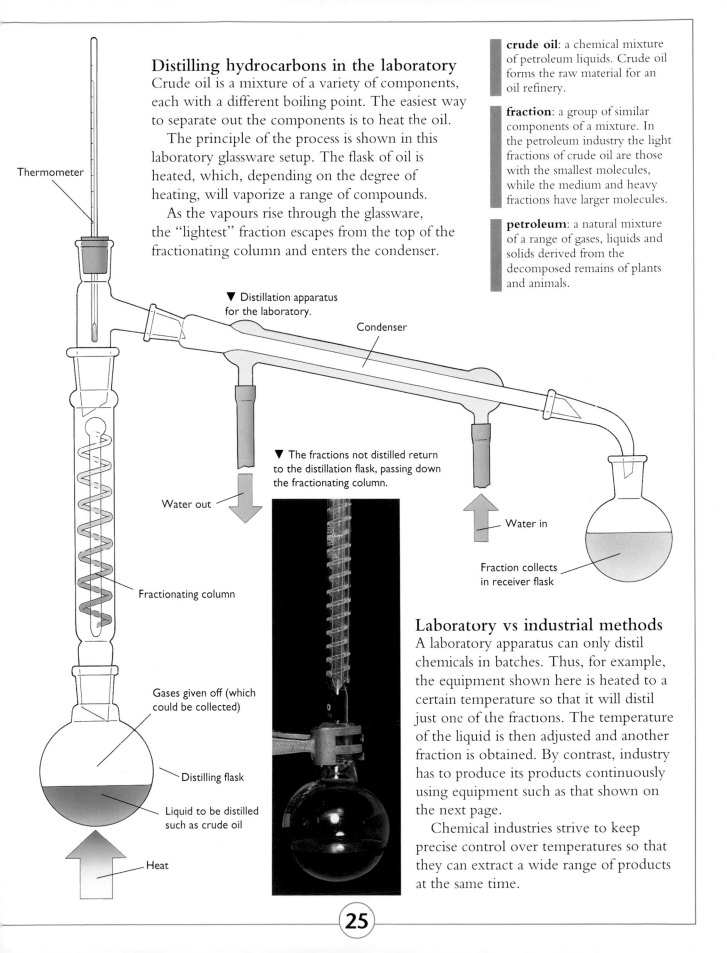

Thermometer

▼ Distillation apparatus for the laboratory.

Condenser

▼ The fractions not distilled return to the distillation flask, passing down the fractionating column.

Water out

Water in

Fraction collects in receiver flask

Fractionating column

Gases given off (which could be collected)

Distilling flask

Liquid to be distilled such as crude oil

Heat

Laboratory vs industrial methods

A laboratory apparatus can only distil chemicals in batches. Thus, for example, the equipment shown here is heated to a certain temperature so that it will distil just one of the fractions. The temperature of the liquid is then adjusted and another fraction is obtained. By contrast, industry has to produce its products continuously using equipment such as that shown on the next page.

Chemical industries strive to keep precise control over temperatures so that they can extract a wide range of products at the same time.

Processing crude oil

Processing crude oil to obtain petrochemicals provides us with some of the most valuable products in use today. This is carbon-chemistry in action. Processing is vital because petroleum is a complicated mixture of chemicals that cannot be used directly. The chemical sorting process is called refining.

In general the most usable fractions are light liquids, such as gasoline, and gases, such as butane. The least desirable are thick liquids. Thus, a petrochemical plant has a second task: to break up, or "crack", the large molecules of the heavy, thick liquid products to make more of those with higher demand.

Cracking is possible because hydrocarbons can be converted from one type to another quite easily. Cracking also has another benefit: if a petrochemical plant did not perform this valuable piece of chemistry, then the amount of waste liquid to be disposed of would be colossal.

Eventually the raw materials for a vast array of complicated products such as plastics, solvents, synthetic fibres, synthetic rubber as well as fuels and solvents can be produced.

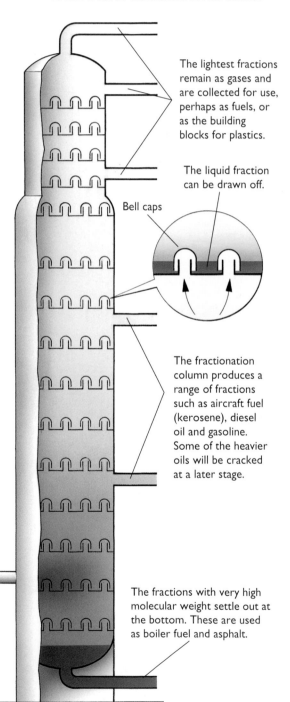

▼ This is a fractionating tower from a refinery and is designed to distil a range of fractions. The fractions with the lowest boiling points condense at the top, whereas those with higher boiling points distil lower down.

At each level small bell caps allow vapours to rise and at the same time trap condensate so that it can be withdrawn from the column.

The lightest fractions remain as gases and are collected for use, perhaps as fuels, or as the building blocks for plastics.

The liquid fraction can be drawn off.

Bell caps

The fractionation column produces a range of fractions such as aircraft fuel (kerosene), diesel oil and gasoline. Some of the heavier oils will be cracked at a later stage.

Heated crude oil enters some way from the bottom.

The fractions with very high molecular weight settle out at the bottom. These are used as boiler fuel and asphalt.

◄ The tall, slim, towers are the fractionating towers of a petrochemical plant. The squat tower on the left is the cooling tower of a power station.

What a petrochemical plant does

A refinery is a collection of tall towers, called distillation towers, connected by a maze of pipes. The crude oil is heated in the towers so that the lighter parts of the mixture boil off, or vapourize. The compounds can be separated out because petroleum is a mixture. This means that the separate fractions can be obtained by making use of the various boiling points of the substances in the mixture.

The gases then condense in the upper, cooler parts of the towers and form into liquids. By using a series of towers, the various chemicals can be separated out. The aim is to produce such different products as liquified petroleum gas (LPG), gasoline (vehicle fuel), kerosene (aircraft fuel), heating oil, diesel fuel, and asphalt (for road surfaces).

A simple process of boiling separates only about one-tenth of the crude oil. The remainder has to be processed again, using a procedure called "cracking", which involves splitting the large, heavy hydrocarbon molecules into smaller ones. A variety of methods are used, including heat and pressure, vacuum, and chemicals such as hydrogen. In this way the products of the refinery can be tailored to market needs. So, for example, if the demand is for an increase in lubricating oils, then the cracking process is adjusted to produce these products; if the oil demand falls and the gasoline demand rises, the type of cracking is changed to give the new balance of products.

cracking: breaking down complex molecules into simpler components. It is a term particularly used in oil refining.

crude oil: a chemical mixture of petroleum liquids. Crude oil forms the raw material for an oil refinery.

fraction: a group of similar components of a mixture. In the petroleum industry the light fractions of crude oil are those with the smallest molecules, while the medium and heavy fractions have larger molecules.

petroleum: a natural mixture of a range of gases, liquids and solids derived from the decomposed remains of plants and animals.

▼ A diagrammatic representation of the cracking process.

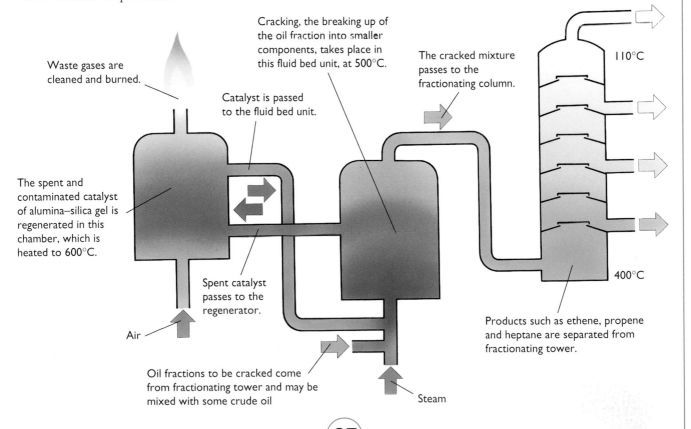

Waste gases are cleaned and burned.

Cracking, the breaking up of the oil fraction into smaller components, takes place in this fluid bed unit, at 500°C.

Catalyst is passed to the fluid bed unit.

The cracked mixture passes to the fractionating column.

110°C

The spent and contaminated catalyst of alumina–silica gel is regenerated in this chamber, which is heated to 600°C.

Spent catalyst passes to the regenerator.

Air

Oil fractions to be cracked come from fractionating tower and may be mixed with some crude oil

Steam

400°C

Products such as ethene, propene and heptane are separated from fractionating tower.

Variety in organic compounds

Carbon chemistry relies heavily on the use of hydrocarbon products obtained from petroleum. On these two pages you can see the structure of some common hydrocarbon substances.

Alkanes (paraffins)

Carbon atoms always form four chemical bonds, that is, they can each bond to four other atoms. All hydrocarbons in which the carbon atoms are joined by a "single" bond have a name which ends in –ane, such as ethane, methane and butane.

▲ Butane gas is used in cooking stoves.

▼ This represents a molecule of ethane. Ethane, butane and pentane gases are obtained from the top of a fractionating column (see page 26).

▼ This represents a molecule of butane.

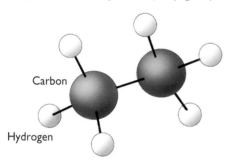

Carbon

Hydrogen

Carbonyl compounds and alcohols

Carbon atoms can use two bonds to join to an oxygen atom, as in the case of the solvent acetone, as well as bonding to oxygen by a single bond as in the alcohol called ethanol.

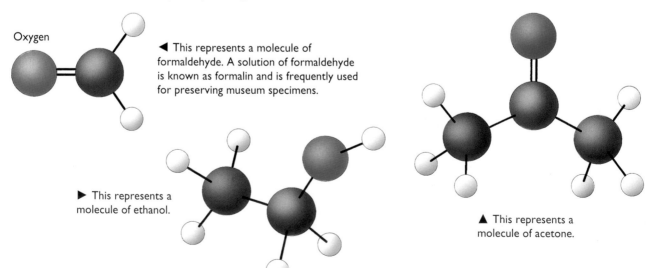

Oxygen

◀ This represents a molecule of formaldehyde. A solution of formaldehyde is known as formalin and is frequently used for preserving museum specimens.

▶ This represents a molecule of ethanol.

▲ This represents a molecule of acetone.

Unsaturated hydrocarbons

Carbon atoms may be joined together with a double bond as in ethene or by a triple bond such as acetylene (whose modern name is ethyne). Acetylene can be used in cutting and welding flames; it requires an oxygen supply if it is to combust completely to form carbon dioxide and water.

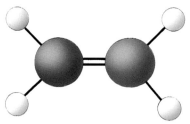

▲ This represents a molecule of ethene.

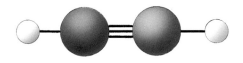

▲ This represents a molecule of ethyne.

Solvents, fats and flavours (esters)

Other groups of atoms found in organic compounds include the "ester" groups, for example ethyl ethanoate (also known as ethyl acetate), which makes nail-varnish remover.

Simple esters are flavouring essences and solvents. Those with a greater number of carbon and hydrogen atoms are cooking oils and fats.

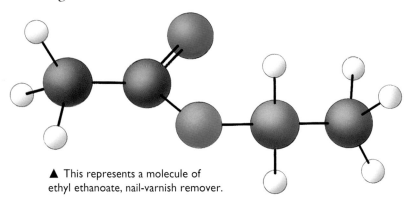

▲ This represents a molecule of ethyl ethanoate, nail-varnish remover.

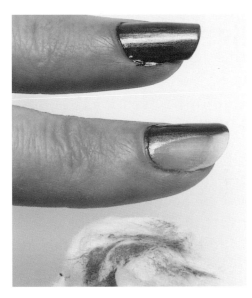

▲ Nail varnish can be removed using a solvent, nail-varnish remover, made from ethyl ethanoate.

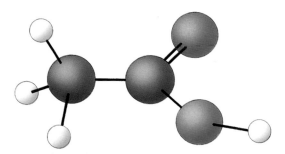

▲ Two oxygen atoms are also present in organic acids such as acetic acid (best known as vinegar).

Also...
Hydrocarbons: the foundation of organic chemistry

The major branch of chemistry that deals with hydrocarbons is often called organic chemistry. This is a historical name, from the mistaken belief that such compounds could only be made by living organisms. Today most of the compounds are made from petroleum in the laboratory or in a petrochemicals plant.

The properties of plastics

Plastics are a group of carbon-based materials that scientists often refer to as polymers. They vary greatly, but they have a number of properties in common.

•Most plastics are not very strong and could not, for example, replace steel in uses such as the frames of buildings.

•Most plastics are easy to bend.

•Many plastics will change shape if they are pulled.

•Many plastics are not very hard and will scratch easily.

•Many plastics are not very dense, that is, they are quite lightweight materials. Some will float in water; most of the others are only just a little denser than water.

•Many plastics are very brittle at low temperatures, that is, they will harden and crack easily in very cold conditions.

•Many plastics soften easily when they get hot (and so cannot be used for things like oven-based cookware). However, there are exceptions, such as nonstick materials.

•Plastics expand about ten times as much as metals as they get hotter.

•Many plastics are destroyed by fire (they decompose), even though they do not actually burn.

•Most plastics are very good electrical insulators, and so can be used to insulate cables, for plugs and sockets, etc.

•Plastics tend to become brittle with age. This is especially so with those exposed to direct sunlight (ultraviolet radiation).

•Plastics are mostly very resistant to being broken down by chemical attack, and they are not water-soluble, which is why they do not decompose when placed out in the open or buried.

Bakelite

Bakelite is a plastic that is based on the substances phenol and formaldehyde. It is a thermosetting plastic, meaning that when heat is applied, it hardens. Bakelite will not catch fire and so was used instead of the inflammable celluloid.

◀ Bakelite appliance from the 1950s.

The history of plastics

The very first plastic, invented over a century ago, was called celluloid. It was a brittle white substance that was used as synthetic ivory. It was also used as the backing to rolls of photographic film.

It took half a century before the next plastic was invented. This was called Bakelite, after its inventor Leo Baekland. But after this, new forms of plastic were invented at a faster and faster rate.

The most well known plastics are nylon and polyester (both widely used for making garments), and polythene and polystyrene (used for making containers and packaging materials).

Most recently there has been a trend to make custom plastics, designed for a specific use.

plastic (material): a carbon-based material consisting of long chains (polymers) of simple molecules. The word plastic is commonly restricted to synthetic polymers.

plastic (property): a material is plastic if it can be made to change shape easily. Plastic materials will remain in the new shape. (Compare with elastic, a property where a material goes back to its original shape.)

polymer: a compound that is made of long chains by combining molecules (called monomers) as repeating units. ("Poly" means many, "mer" means part).

Celluloid: the first plastic

Alexander Parkes made the first plastic in 1856. It is made from cellulose nitrate and camphor. Many nitrate compounds are used in explosives, and celluloid material was used in World War 1 as a smokeless explosive. Not surprisingly, one of the problems with celluloid in the home or office is that it catches fire very easily. As a result it is rarely used today.

On the other hand cellulose, which is also prone to catch fire, is still widely used as a solvent for inks and paints because it dries very quickly. Car body paints, for example, are formulated with cellulose.

▶ Celluloid table-tennis balls

Nonstick polymers

Nonstick surfaces on many pots and pans are made from polytetrafluoroethene (Teflon), which is based on the two simple chemicals fluorine and ethene. It was invented in 1938. It does not dissolve or burn. This is why it can be used on pans and why fats, oils and other materials used in cooking have no effect on it.

◀ A modern nonstick frying pan with a Teflon surface.

Also... Plastics solvents

A solvent is a liquid chemical compound used for dissolving another compound without reacting with it. Most organic materials do not dissolve in water, so water cannot be used as a solvent.

Most solvents have a low boiling point and are volatile, that is, they evaporate easily. A solvent used in polymer-based paints is cellulose nitrate; a solvent for nail-varnish (cellulose acetate) is ethyl ethanoate (see page 29).

Making polymers

Polymers are made in two ways: by adding building blocks together in long chains to produce materials known as addition polymers, and by removing water molecules to produce materials known as condensation polymers. These two types of polymer are important, and they produce such a wide range of everyday materials. On the next few pages is information on addition polymers; condensation polymers are described on page 36. The remainder of the book will look at some important polymers.

Addition polymerisation

Some hydrocarbon molecules are very reactive and can be joined together, or polymerised. The reactive units are those in which carbon atoms are joined by a double bond. Ethene, shown below, is an example.

To polymerise these units, or monomers, the double bond is broken and one bond used to join the unit to its neighbours. This builds long chains and is known as "addition polymerisation".

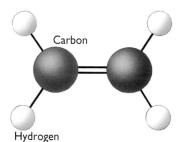

Carbon

Hydrogen

◀ This represents a molecule of ethene. Units like this that can be polymerised are known as monomers.

▶ This represents a unit of addition polymerisation derived from ethene. The ethene molecules join together to form the polymer polyethene, more commonly known as polythene.

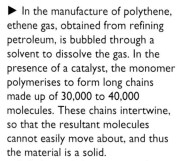

▶ In the manufacture of polythene, ethene gas, obtained from refining petroleum, is bubbled through a solvent to dissolve the gas. In the presence of a catalyst, the monomer polymerises to form long chains made up of 30,000 to 40,000 molecules. These chains intertwine, so that the resultant molecules cannot easily move about, and thus the material is a solid.

Polythene is cheap to manufacture and is used to make polythene bags and buckets. The properties of this polymer can be altered by substituting some or all of the hydrogens in the hydrocarbon chain with other elements or compounds. Some of these other polymers are shown on the opposite page.

A great range of polymers

Other monomers that produce addition polymers include tetrafluoroethene, chloroethene and styrene. These are shown below.

► In the case of tetrafluoroethene all the hydrogen atoms on the ethene molecule have been substituted by fluorine. The polymer is polytetrafluoroethene (PTFE or Teflon), a hard plastic which is not attacked by most chemicals. It is used on such items as nonstick pans.

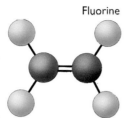
Fluorine

► In chloroethene (vinyl chloride), a hydrogen atom has been substituted by chlorine. Its polymer is polychloroethene (polyvinyl chloride or PVC).

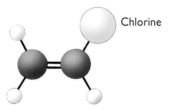
Chlorine

▼ In styrene a hydrogen atom on the ethene molecule has been replaced by a ring of carbon atoms, known as a benzene ring. Styrene is polymerised to produce polystyrene.

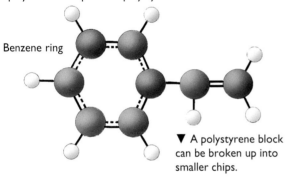

Benzene ring

▼ A polystyrene block can be broken up into smaller chips.

Polystyrene

Polystyrene (polyvinyl benzene), or Styrofoam as it is often called, is a hard, brittle plastic made by blowing air into the softened plastic. It is frequently used for insulation and its light weight makes it ideal for packaging. However, it is very brittle, and cannot withstand knocks as well as other plastics such as PVC.

Polystyrene is very hazardous in a fire because as it melts, molten globules flow away, carrying the fire with them and thus helping spread it.

catalyst: a substance that speeds up a chemical reaction but itself remains unaltered at the end of the reaction.

monomer: a building block of a larger chain molecule ("mono" means one, "mer" means part).

oxidise: the process of gaining oxygen. This can be part of a controlled chemical reaction, or it can be the result of exposing a substance to the air, where oxidation (a form of corrosion) will occur slowly, perhaps over months or years.

polymer: a compound that is made of long chains by combining molecules (called monomers) as repeating units. ("Poly" means many, "mer" means part).

thermoplastic: a plastic that will soften, can repeatedly be moulded it into shape on heating and will set into the moulded shape as it cools.

Vinyl (PVC)

Vinyl, or PVC (polyvinyl chloride), is a form of plastic used widely as a furniture covering, in moulded items such as washing up bowls, electrical cable insulation, window frames, house panels and for many other items.

It is a thermosetting plastic, meaning that the objects have to be formed from the polymer while it is still a hot liquid. Once it has cooled and set it will retain the shape that it has been given and will not soften again.

Vinyl will harden and crack when exposed to ultraviolet light (a natural component of sunlight). It will also oxidise in air. To make it stable, give it attractive colours, and make it resistant to knocks, other materials have to be combined with it. Carbon powder, for example, is added to make the black-coloured plastic that can be used for waste bags and other uses where the material will be exposed.

▲ This house cladding and the window frames are made from u-PVC, which has been specially treated to prevent deterioration and discolouration when subject to ultraviolet light. This means that the house doesn't need to be painted, as similar, traditionally built houses would.

Rubber, a natural addition polymer

Rubber is the name of an elastic material that can be stretched to several times its own length before breaking. The word rubber originally referred to the material, known as latex, that is the sap of the rubber tree. However, a wide variety of synthetic rubbers are now also produced.

The rubber molecule is an example of a polymer, a giant molecule made of tens of thousands to millions of simple units arranged in a chain. The individual units (called monomers) are each about the size of a sugar molecule.

The rubber polymer is elastic because the long chains are linked together (called cross-linking). Each chain can be pulled slightly past each other chain. The cross-linking, however, pulls the chains back into their original place when the force is removed.

Rubbers can be more strongly cross-linked using sulphur through a process called vulcanisation. Vulcanisation produces the very strong form of rubber used for vehicle tyres.

▲◄ Latex is tapped from rubber trees and then collected for processing. It is still a very labour-intensive operation.

▲▼ Rubber is used in a very wide range of products wherever elastic, waterproof properties are required. In this factory rubber is being used for making gloves for laboratory use. The formers, made of a ceramic material, are dipped into a bath of latex, giving them a coating. The coated formers are transported through an oven, where the rubber dries. The gloves are pulled off the formers and a drying powder is puffed into them to make them easier to pull on and off. Air is blown into each glove to test for defects. All the gloves used by the chemists in the demonstrations in this series of books were made on this Malaysian production line.

latex: (the Latin word for "liquid") a suspension of small polymer particles in water. The rubber that flows from a rubber tree is a natural latex. Some synthetic polymers are made as latexes, allowing polymerisation to take place in water.

polymerisation: a chemical reaction in which large numbers of similar molecules arrange themselves into large molecules, usually long chains. This process usually happens when there is a suitable catalyst present. For example, ethene reacts to form polythene in the presence of certain catalysts.

vulcanisation: forming cross-links between polymer chains to increase the strength of the whole polymer. Rubbers are vulcanised using sulphur when making tyres and other strong materials.

Also... The history of rubber

Rubber was first noticed by Europeans when Christopher Columbus arrived in the Americas. Natural rubber is a suspension of about 30% rubber particles in water. It is a white latex sap that flows naturally from rubber trees if the bark is scored.

The first person to find a use for the material was Charles Macintosh who, in 1823, discovered that it could be used to make fabric waterproof. However, because natural rubber has no cross-links, it is difficult to use, becoming sticky when hot and stiff when cold. In 1839 Charles Goodyear invented the technique of vulcanising rubber by heating it with sulphur.

As early as 1826, Michael Faraday found that rubber was a polymer of the monomer isoprene. The first artificial rubber was made during World War II, when there was a shortage of natural rubber. Styrene and butadiene are now used as the foundation of the modern synthetic rubber industry because they are more easily obtained than isoprene. They are all derived from petroleum.

▶ Styrene, $C_6H_5CHCH_2$ is a hydrocarbon monomer. It is stored as a liquid, but on exposure to air it polymerises, setting into the shape of the container. In much polymer chemistry, the polymer is formulated so that it has useful properties when solid.

Condensation polymers

The alternative to addition polymerisation is called condensation polymerisation. In this process, each stage in the polymer chain forms as a water molecule is expelled.

Many common fibres are formed by condensation polymerisation, including nylon and polyester. Their structures are shown here and the way they are made is shown on the following pages.

Nylon

Organic compounds containing nitrogen are important substances. Hexan-1,6-diamine can be polymerised with hexandioic acid to form a nylon. This is an example of condensation polymerisation, where a water molecule is eliminated as every link in the polymer chain is formed.

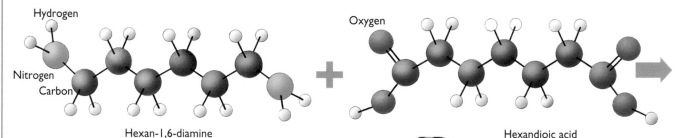

Hexan-1,6-diamine

Hexandioic acid

▶ Proteins form the building blocks of the tissues in animals.

Amino acids

Amino acids (one of the building blocks of proteins) can link in a similar way to form long chains of proteins in order to build tissues. They are natural condensation polymers.

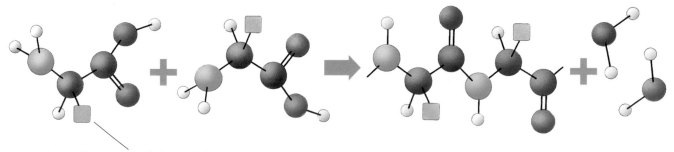

In the simplest amino acid, the shaded box, would be a hydrogen atom. This section of the molecule is different for different amino acids.

A section of a protein chain

Water

amino acid: amino acids are organic compounds that are the building blocks for the proteins in the body.

condensation polymerisation: where a water molecule is eliminated as every link in the polymer chain is formed.

ester: organic compounds, formed by the reaction of an alcohol with an acid, which often have a fruity taste.

◄ Synthetic fibres such as nylon and polyester are cheap to manufacture and are so are used for most modern clothing.

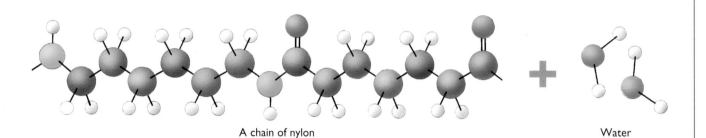

A chain of nylon

Water

Polyesters

The polyester Terylene is produced by heating a mixture of terephthalic acid and ethylene glycol.

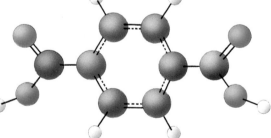

Terephthalic acid

Ethylene glycol

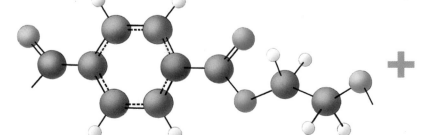

Section of a Terylene polymer chain

Water

Synthetic fibres

In the past, fibres were made from natural materials such as flax, wool and cotton; however, synthetic fibres are made from chemicals. The advantage of synthetic fibres over natural fibres is that the chemist can control the nature of the material closely to suit a specific purpose. On the other hand, natural fibres are so sophisticated that they will often perform functions that chemists do not yet know how to copy synthetically. Even so, synthetic fibres are very attractive because of their flexibility and strength. Many are stronger than metals.

The first fibre was rayon, a treated form of natural plant cellulose. The first entirely synthetic fibre was made in 1938. It was called nylon, because it was due to collaboration between teams in New York (NY) and London (Lon). Like all fibre polymers, nylon is made into a syrup-like substance and then extruded through tiny holes. The filaments (fibres) are then pulled out and wound on to a drum.

Synthetic fibres can be woven into cloth as mixtures to add durability to the material. Many coats, for example, are a mixture of polyester (which is hard wearing) and wool (which is warm and pleasant to the touch).

Synthetic fibres can also be incorporated into resins and made into composite materials with great strength.

Making nylon in the laboratory

A wide variety of plastics can be produced with simple techniques in the laboratory. The two chemicals used here are the organic chemicals hexan-1,6-diamine and hexan-dioyl chloride.

As one liquid is poured into the other, the chemical reaction immediately produces a white solid. This is nylon.

❶▼ Hexan-1,6-diamine is poured carefully on to hexan-dioyl chloride. Here you can see the nylon forming at the interface between the two liquids.

❷▼ The technique of spinning the nylon is used to produce a thread. To do this a piece of the nylon is picked out of the solution using tongs and is carefully lifted clear. Here you can see the difficulty of obtaining a filament with an even diameter. To achieve this the material needs to be lifted from the beaker at a constant rate.

polymer: a compound that is made of long chains by combining molecules (called monomers) as repeating units. ("Poly" means many, "mer" means part).

solvent: the main substance in a solution (e.g. water in salt water).

❸▼ The filament begins to form. Here you can see that the filament can only be produced as long as the two liquids remain. As soon as the batch is used up, the process of picking a filament has to be repeated. Clearly, while the laboratory process shows that it is possible to produce filaments easily, industrial processes have to be found that make a more reliable filament in a continuous process. To do this the threads are extruded, not drawn, as shown on the next page.

Polyester fibre and film

Polyester is a carbon-chain material that is widely used as a synthetic fibre, and it can also be stretched out to make a film. It is not a natural filament as, for example, cotton or wool. Instead, it is pushed out, or extruded from a liquid through find holes in a machine (called a spinneret) and then pulled into long filaments.

Polyester is a synthetic fabric made from woven filaments. It has been designed to be strong (it will not break easily when being woven on a machine), to stand up to many machine washes and to be easy to dye, so that it can be made into many colours. It will also stretch, and so be more comfortable for the wearer. The filaments can then be blended with natural fibres (to give more wear resistance to natural materials) or used on their own.

Some special forms of polyester fibres are given a hollow cross-section. This gives them the property of being good at insulating yet lightweight, so they can be used for warm clothing. Another form of polyester is made of microfibres, which make a mesh that keeps wind and water out, yet allows body moisture to seep through.

▲▼ To make fibres, liquid polymer is extruded. The polymer typically arrives in the form of pellets, which are then heated to make a liquid. The liquid is then extruded through fine nozzles in the spinneret so that filaments (single strands) of plastic emerge. These are cooled and rapidly solidify so they can be twisted together to make a fibre.

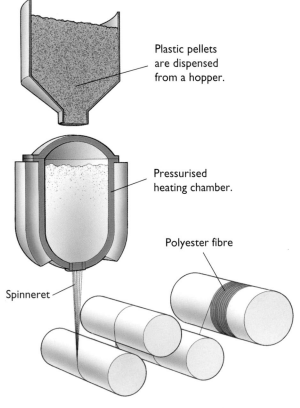

Plastic pellets are dispensed from a hopper.

Pressurised heating chamber.

Polyester fibre

Spinneret

▶ Hollow and microfibre polyester is widely used in clothing. In the case of sportswear the materials have to be waterproof and yet allow body moisture out. They have to resist the scuffing that comes from a fall, and they have to be lightweight so they do not unduly restrict movement.

▼ Filaments of polyester fibre
are wound on to drums in a
modern synthetic fibre factory.

extrusion: forming a shape by pushing it through a die. For example, toothpaste is extruded through the cap (die) of the toothpaste tube.

synthetic: does not occur naturally, but has to be manufactured.

Also... Esters

The word "ester" refers to a group of organic compounds that includes many flavourings and also common solvents and plastics such as acetate (used as a backing on photographic film, for example). Esters are often made by reacting acids with alcohols.

Polyesters are polymers made with ester groups. Polymers used for coatings are made so that the polymers are cross-linked (as are rubbers). The form of polyester used for fibres is called polythene–terephthalate.

Polyesters can also be stretched into films such as the base of photographic films (when they are usually referred to simply as "acetate"). The sails on boats are also usually made from polyesters because of their strength and resistance to rotting.

▲ This photographic film is made from an ester, normally called acetate.

Carbon-based chemicals and the environment

Carbon is found in nine-tenths of all the known compounds. In gases, it has also been used in making more synthetic materials than any other element. People are more concerned about the effects of synthetic materials because it is not always known how they react with people, animals and plants.

People are less concerned about the use of inorganic carbon-based compounds, such as limestone, than they are about organic carbon products, such as oil and coal. For example, the extraction of limestone from a quarry produces a limited environmental impact that can be understood and evaluated easily. The extraction of organic carbon-based chemicals produces much more complex problems. For example, some are made from forest products, and this can have grave implications for forest renewal. Many products are burned, adding to the Greenhouse Effect (see page 12). Oil is extracted from the ground and transported as a liquid, with the accompanying danger of possible oil spills.

Organic products cause concern because of the way they may interact with people and other living things. Many are designed as weedkillers, pesticides and so on, so it is possible that others produced for different purposes could have harmful side effects on living things. All this causes continuing concern, which is why regulation of the organic chemicals industry is so vital.

▲ Burning fossil fuels produces carbon dioxide in the air, adding to the Greenhouse Effect.

Pesticides and herbicides

Some carbon-based dyes and pesticides have been found to be very hazardous to health. The pesticide DDT, for example, was found to be very effective and was used all over the world. Years later, however, it was found to be very harmful to fish and birds and its use is now banned in most parts of the world.

Oil

The main area of concern involves oil spills, because the thick layers of oil take time to be oxidised to carbon dioxide gas, or to be digested by organisms in water. Before natural processes can react with the oil, many animals are likely to die from ingesting the oil.

▶ Clearing up after the *Exxon Valdez* spill in Prince William Sound, Alaska.

Plastics

Another area of concern is in the use of plastics. Because these materials are inert, that is, they do not oxidise, they are insoluble in water and cannot be digested by the organisms that normally cause decay.

Special formulations, used as substitutes for petroleum-based plastics and based on starches from plants may help solve some of these problems for plastics used in packaging. These polymers decompose naturally in the wet conditions of the soil and so will decay in landfill sites.

oxidise: the process of gaining oxygen. This can be part of a controlled chemical reaction, or it can be the result of exposing a substance to the air, where oxidation (a form of corrosion) will occur slowly, perhaps over months or years.

Organic chemical releases

Sometimes fumes emanating from chemical works can have disastrous effects, as they did in Bhopal, India, in 1984. Here a pesticide plant accidentally allowed water into a tank of methyl isocyanate. The result was the release of a toxic gas that killed 2500 people and injured hundreds of thousands more.

▲ A polystyrene cup that will not decay.

▶ Refuse collection sacks are now being designed to decompose, as is some packaging.

43

Key facts about...

Carbon

The sixth most common element, making 0.2% of the Earth's crust

A non-metal, chemical symbol C

The most versatile of all the elements; 94% of all compounds contain carbon

An important building block of all natural substances

Hardest form is diamond

Carbon dioxide is an important gas in the air

Can be burned as a fuel

Can combine into long chains, the basis of plastics

Good conductor of electricity

Melts at a high temperature (3550°C)

Atomic number 6, atomic weight about 12

Used as the standard from which the atomic weights of all other elements are calculated

SHELL DIAGRAMS

The shell diagram on this page represents an atom of the element carbon. The total number of electrons are shown in the relevant orbitals, or shells, around the central nucleus.

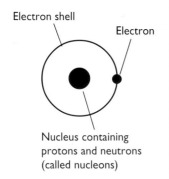

Electron shell

Electron

Nucleus containing protons and neutrons (called nucleons)

▶ Engine oil is one of the many fractions of crude oil extracted by fractionation in a refinery.

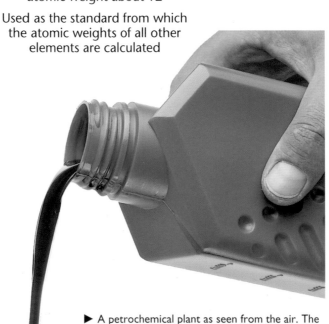

▶ A petrochemical plant as seen from the air. The organisation of such a plant is a very complex task. Each of the processes has to be arranged so that the products from one plant become the raw materials for another. Many materials also have to be recycled so they can be processed properly. This adds to the complexity of the site.

The Periodic Table

The Periodic Table sets out the relationships among the elements of the Universe. According to the Periodic Table, certain elements fall into groups. The pattern of these groups has, in the past, allowed scientists to predict elements that had not at that time been discovered. It can still be used today to predict the properties of unfamiliar elements.

The Periodic Table was first described by a Russian teacher, Dmitry Ivanovich Mendeleev, between 1869 and 1870. He was interested in writing a chemistry textbook, and wanted to show his students that there were certain patterns in the elements that had been discovered. So he set out the elements (of which there were 57 at the time) according to their known properties. On the assumption that there was pattern to the elements, he left blank spaces where elements seemed to be missing. Using this first version of the Periodic Table, he was able to predict in detail the chemical and physical properties of elements that had not yet been discovered. Other scientists began to look for the missing elements, and they soon found them.

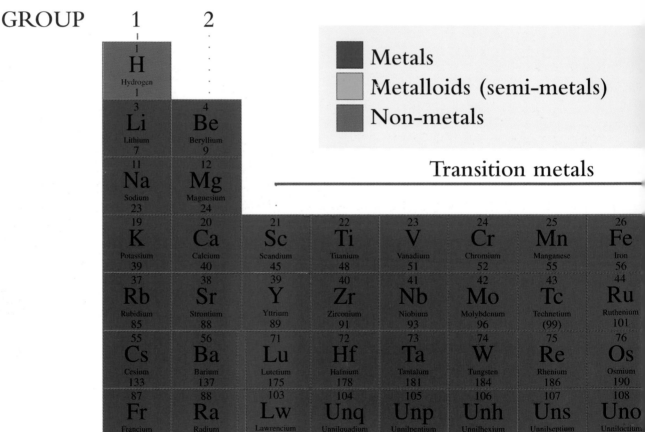

GROUP

Metals
Metalloids (semi-metals)
Non-metals

Transition metals

Lanthanide metals

Actinoid metals

Hydrogen did not seem to fit into the table, so he placed it in a box on its own. Otherwise the elements were all placed horizontally. When an element was reached with properties similar to the first one in the top row, a second row was started. By following this rule, similarities among the elements can be found by reading up and down. By reading across the rows, the elements progressively increase their atomic number. This number indicates the number of positively charged particles (protons) in the nucleus of each atom. This is also the number of negatively charged particles (electrons) in the atom.

The chemical properties of an element depend on the number of electrons in the outermost shell.

Atoms can form compounds by sharing electrons in their outermost shells. This explains why atoms with a full set of electrons (like helium, an inert gas) are unreactive, whereas atoms with an incomplete electron shell (such as chlorine) are very reactive. Elements can also combine by the complete transfer of electrons from metals to non-metals and the compounds formed contain ions.

Radioactive elements lose particles from their nucleus and electrons from their surrounding shells. As a result their atomic number changes and they become new elements.

Atomic (proton) number

13
Al
Aluminium
27

Symbol

Name

Approximate relative atomic mass
(Approximate atomic weight)

3	4	5	6	7	0
					2 He Helium 4
5 B Boron 11	6 C Carbon 12	7 N Nitrogen 14	8 O Oxygen 16	9 F Fluorine 19	10 Ne Neon 20
13 Al Aluminium 27	14 Si Silicon 28	15 P Phosphorus 31	16 S Sulphur 32	17 Cl Chlorine 35	18 Ar Argon 40

27 Co Cobalt 59	28 Ni Nickel 59	29 Cu Copper 64	30 Zn Zinc 65	31 Ga Gallium 70	32 Ge Germanium 73	33 As Arsenic 75	34 Se Selenium 79	35 Br Bromine 80	36 Kr Krypton 84
45 Rh Rhodium 103	46 Pd Palladium 106	47 Ag Silver 108	48 Cd Cadmium 112	49 In Indium 115	50 Sn Tin 119	51 Sb Antimony 122	52 Te Tellurium 128	53 I Iodine 127	54 Xe Xenon 131
77 Ir Iridium 192	78 Pt Platinum 195	79 Au Gold 197	80 Hg Mercury 201	81 Tl Thallium 204	82 Pb Lead 207	83 Bi Bismuth 209	84 Po Polonium (209)	85 At Astatine (210)	86 Rn Radon (222)

109
Une
Unnilennium
(266)

61 Pm Promethium (145)	62 Sm Samarium 150	63 Eu Europium 152	64 Gd Gadolinium 157	65 Tb Terbium 159	66 Dy Dysprosium 163	67 Ho Holmium 165	68 Er Erbium 167	69 Tm Thulium 169	70 Yb Ytterbium 173
93 Np Neptunium (237)	94 Pu Plutonium (244)	95 Am Americium (243)	96 Cm Curium (247)	97 Bk Berkelium (247)	98 Cf Californium (251)	99 Es Einsteinium (252)	100 Fm Fermium (257)	101 Md Mendelevium (258)	102 No Nobelium (259)

Understanding equations

As you read through this book, you will notice that many pages contain equations using symbols. If you are not familiar with these symbols, read this page. Symbols make it easy for chemists to write out the reactions that are occurring in a way that allows a better understanding of the processes involved.

Symbols for the elements

The basis of the modern use of symbols for elements dates back to the 19th century. At this time a shorthand was developed using the first letter of the element wherever possible. Thus "O" stands for oxygen, "H" stands for hydrogen

and so on. However, if we were to use only the first letter, then there could be some confusion. For example, nitrogen and nickel would both use the symbols N. To overcome this problem, many elements are symbolised using the first two letters of their full name, and the second letter is lowercase. Thus although nitrogen is N, nickel becomes Ni. Not all symbols come from the English name; many use the Latin name instead. This is why, for example, gold is not G but Au (for the Latin *aurum*) and sodium has the symbol Na, from the Latin *natrium*.

Compounds of elements are made by combining letters. Thus the molecule carbon

Written and symbolic equations

In this book, important chemical equations are briefly stated in words (these are called word equations), and are then shown in their symbolic form along with the states.

What reaction the equation illustrates

EQUATION: The formation of calcium hydroxide

Word equation ——— *Calcium oxide + water ⇨ calcium hydroxide*

Symbol equation ——— $CaO(s)$ $+$ $H_2O(l)$ ⇨ $Ca(OH)_2(aq)$

heated

Sometimes you will find additional descriptions below the symbolic equation.

Symbol showing the state: s is for solid, l is for liquid, g is for gas and aq is for aqueous.

Diagrams

Some of the equations are shown as graphic representations.

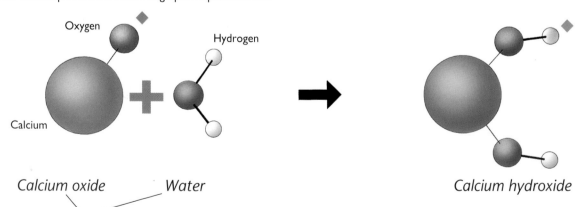

Oxygen

Hydrogen

Calcium

Calcium oxide   *Water*

Calcium hydroxide

Sometimes the written equation is broken up and put below the relevant stages in the graphic representation.

monoxide is CO. By using lowercase letters for the second letter of an element, it is possible to show that cobalt, symbol Co, is not the same as the molecule carbon monoxide, CO.

However, the letters can be made to do much more than this. In many molecules, atoms combine in unequal numbers. So, for example, carbon dioxide has one atom of carbon for every two of oxygen. This is shown by using the number 2 beside the oxygen, and the symbol becomes CO_2.

In practice, some groups of atoms combine as a unit with other substances. Thus, for example, calcium bicarbonate (one of the compounds used in some antacid pills) is written $Ca(HCO_3)_2$. This shows that the part of the substance inside the brackets reacts as a unit and the "2" outside the brackets shows the presence of two such units.

Some substances attract water molecules to themselves. To show this a dot is used. Thus the blue form of copper sulphate is written $CuSO_4.5H_2O$. In this case five molecules of water attract to one of copper sulphate.

When you see the dot, you know that this water can be driven off by heating; it is part of the crystal structure.

In a reaction substances change by rearranging the combinations of atoms. The way they change is shown by using the chemical symbols, placing those that will react (the starting materials, or reactants) on the left and the products of the reaction on the right. Between the two, chemists use an arrow to show which way the reaction is occurring.

It is possible to describe a reaction in words. This gives word equations, which are given throughout this book. However, it is easier to understand what is happening by using an equation containing symbols. These are also given in many places. They are not given when the equations are very complex.

In any equation both sides balance; that is, there must be an equal number of like atoms on both sides of the arrow. When you try to write down reactions, you, too, must balance your equation; you cannot have a few atoms left over at the end!

The symbols in brackets are abbreviations for the physical state of each substance taking part, so that (s) is used for solid, (l) for liquid, (g) for gas and (aq) for an aqueous solution, that is, a solution of a substance dissolved in water.

Atoms and ions
Each sphere represents a particle of an element. A particle can be an atom or an ion. Each atom or ion is associated with other atoms or ions through bonds – forces of attraction. The size of the particles and the nature of the bonds can be extremely important in determining the nature of the reaction or the properties of the compound.

Chemical symbols, equations and diagrams
The arrangement of any molecule or compound can be shown in one of the two ways shown below, depending on which gives the clearer picture. The left-hand diagram is called a ball-and-stick diagram because it uses rods and spheres to show the structure of the material. This example shows water, H_2O. There are two hydrogen atoms and one oxygen atom.

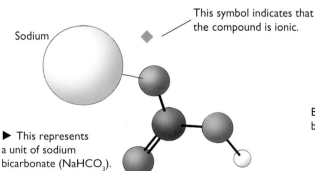

Sodium

This symbol indicates that the compound is ionic.

▶ This represents a unit of sodium bicarbonate ($NaHCO_3$).

The term "unit" is sometimes used to simplify the representation of a combination of ions.

Bond shown by "stick"

Colours too
The colours of each of the particles help differentiate the elements involved. The diagram can then be matched to the written and symbolic equation given with the diagram. In the case above, oxygen is red and hydrogen is grey.

Glossary of technical terms

absorb: to soak up a substance. Compare to adsorb.

acetone: a petroleum-based solvent.

acid: compounds containing hydrogen which can attack and dissolve many substances. Acids are described as weak or strong, dilute or concentrated, mineral or organic.

acidity: a general term for the strength of an acid in a solution.

acid rain: rain that is contaminated by acid gases such as sulphur dioxide and nitrogen oxides released by pollution.

adsorb/adsorption: to "collect" gas molecules or other particles on to the *surface* of a substance. They are not chemically combined and can be removed. (The process is called "adsorption".) Compare to absorb.

alchemy: the traditional "art" of working with chemicals that prevailed through the Middle Ages. One of the main challenges of alchemy was to make gold from lead. Alchemy faded away as scientific chemistry was developed in the 17th century.

alkali: a base in solution.

alkaline: the opposite of acidic. Alkalis are bases that dissolve, and alkaline materials are called basic materials. Solutions of alkalis have a pH greater than 7.0 because they contain relatively few hydrogen ions.

alloy: a mixture of a metal and various other elements.

alpha particle: a stable combination of two protons and two neutrons, which is ejected from the nucleus of a radioactive atom as it decays. An alpha particle is also the nucleus of the atom of helium. If it captures two electrons it can become a neutral helium atom.

amalgam: a liquid alloy of mercury with another metal.

amino acid: amino acids are organic compounds that are the building blocks for the proteins in the body.

amorphous: a solid in which the atoms are not arranged regularly (i.e. "glassy"). Compare with crystalline.

amphoteric: a metal that will react with both acids and alkalis.

anhydrous: a substance from which water has been removed by heating. Many hydrated salts are crystalline. When they are heated and the water is driven off, the material changes to an anhydrous powder.

anion: a negatively charged atom or group of atoms.

anode: the negative terminal of a battery or the positive electrode of an electrolysis cell.

anodising: a process that uses the effect of electrolysis to make a surface corrosion-resistant.

antacid: a common name for any compound that reacts with stomach acid to neutralise it.

antioxidant: a substance that prevents oxidation of some other substance.

aqueous: a solid dissolved in water. Usually used as "aqueous solution".

atom: the smallest particle of an element.

atomic number: the number of electrons or the number of protons in an atom.

atomised: broken up into a very fine mist. The term is used in connection with sprays and engine fuel systems.

aurora: the "northern lights" and "southern lights" that show as coloured bands of light in the night sky at high latitudes. They are associated with the way cosmic rays interact with oxygen and nitrogen in the air.

basalt: an igneous rock with a low proportion of silica (usually below 55%). It has microscopically small crystals.

base: a compound that may be soapy to the touch and that can react with an acid in water to form a salt and water.

battery: a series of electrochemical cells.

bauxite: an ore of aluminium, of which about half is aluminium oxide.

becquerel: a unit of radiation equal to one nuclear disintegration per second.

beta particle: a form of radiation in which electrons are emitted from an atom as the nucleus breaks down.

bleach: a substance that removes stains from materials either by oxidising or reducing the staining compound.

boiling point: the temperature at which a liquid boils, changing from a liquid to a gas.

bond: chemical bonding is either a transfer or sharing of electrons by two or more atoms. There are a number of types of chemical bond, some very strong (such as covalent bonds), others weak (such as hydrogen bonds). Chemical bonds form because the linked molecule is more stable than the unlinked atoms from which it formed. For example, the hydrogen molecule (H_2) is more stable than single atoms of hydrogen, which is why hydrogen gas is always found as molecules of two hydrogen atoms.

brass: a metal alloy principally of copper and zinc.

brazing: a form of soldering, in which brass is used as the joining metal.

brine: a solution of salt (sodium chloride) in water.

bronze: an alloy principally of copper and tin.

buffer: a chemistry term meaning a mixture of substances in solution that resists a change in the acidity or alkalinity of the solution.

capillary action: the tendency of a liquid to be sucked into small spaces, such as between objects and through narrow-pore tubes. The force to do this comes from surface tension.

catalyst: a substance that speeds up a chemical reaction but itself remains unaltered at the end of the reaction.

cathode: the positive terminal of a battery or the negative electrode of an electrolysis cell.

cathodic protection: the technique of making the object that is to be protected from corrosion into the cathode of a cell. For example, a material, such as steel, is protected by coupling it with a more reactive metal, such as magnesium. Steel forms the cathode and magnesium the anode. Zinc protects steel in the same way.

cation: a positively charged atom or group of atoms.

caustic: a substance that can cause burns if it touches the skin.

cell: a vessel containing two electrodes and an electrolyte that can act as an electrical conductor.

ceramic: a material based on clay minerals, which has been heated so that it has chemically hardened.

chalk: a pure form of calcium carbonate made of the crushed bodies of microscopic sea creatures, such as plankton and algae.

change of state: a change between one of the three states of matter, solid, liquid and gas.

chlorination: adding chlorine to a substance.

cladding: a surface sheet of material designed to protect other materials from corrosion.

clay: a microscopically small plate-like mineral that makes up the bulk of many soils. It has a sticky feel when wet.

combustion: the special case of oxidisation of a substance where a considerable amount of heat and usually light are given out. Combustion is often referred to as "burning".

compound: a chemical consisting of two or more elements chemically bonded together. Calcium atoms can combine with carbon atoms and oxygen atoms to make calcium carbonate, a compound of all three atoms.

condensation nuclei: microscopic particles of dust, salt and other materials suspended in the air, which attract water molecules.

conduction: (i) the exchange of heat (heat conduction) by contact with another object or (ii) allowing the flow of electrons (electrical conduction).

convection: the exchange of heat energy with the surroundings produced by the flow of a fluid due to being heated or cooled.

corrosion: the *slow* decay of a substance resulting from contact with gases and liquids in the environment. The term is often applied to metals. Rust is the corrosion of iron.

corrosive: a substance, either an acid or an alkali, that *rapidly* attacks a wide range of other substances.

cosmic rays: particles that fly through space and bombard all atoms on the Earth's surface. When they interact with the atmosphere they produce showers of secondary particles.

covalent bond: the most common form of strong chemical bonding, which occurs when two atoms *share* electrons.

cracking: breaking down complex molecules into simpler components. It is a term particularly used in oil refining.

crude oil: a chemical mixture of petroleum liquids. Crude oil forms the raw material for an oil refinery.

crystal: a substance that has grown freely so that it can develop external faces. Compare with crystalline, where the atoms are not free to form individual crystals and amorphous where the atoms are arranged irregularly.

crystalline: the organisation of atoms into a rigid "honeycomb-like" pattern without distinct crystal faces.

crystal systems: seven patterns or systems into which all of the world's crystals can be grouped. They are: cubic, hexagonal, rhombohedral, tetragonal, orthorhombic, monoclinic and triclinic.

cubic crystal system: groupings of crystals that look like cubes.

curie: a unit of radiation. The amount of radiation emitted by 1 g of radium each second. (The curie is equal to 37 billion becquerels.)

current: an electric current is produced by a flow of electrons through a conducting solid or ions through a conducting liquid.

decay (radioactive decay): the way that a radioactive element changes into another element because of loss of mass through radiation. For example uranium decays (changes) to lead.

decompose: to break down a substance (for example by heat or with the aid of a catalyst) into simpler components. In such a chemical reaction only one substance is involved.

dehydration: the removal of water from a substance by heating it, placing it in a dry atmosphere, or through the action of a drying agent.

density: the mass per unit volume (e.g. g/cc).

desertification: a process whereby a soil is allowed to become degraded to a state in which crops can no longer grow, i.e. desert-like. Chemical desertification is usually the result of contamination with halides because of poor irrigation practices.

detergent: a petroleum-based chemical that removes dirt.

diaphragm: a semipermeable membrane – a kind of ultra-fine mesh filter – that will allow only small ions to pass through. It is used in the electrolysis of brine.

diffusion: the slow mixing of one substance with another until the two substances are evenly mixed.

digestive tract: the system of the body that forms the pathway for food and its waste products. It begins at the mouth and includes the stomach and the intestines.

dilute acid: an acid whose concentration has been reduced by a large proportion of water.

diode: a semiconducting device that allows an electric current to flow in only one direction.

disinfectant: a chemical that kills bacteria and other microorganisms.

dissociate: to break apart. In the case of acids it means to break up forming hydrogen ions. This is an example of ionisation. Strong acids dissociate completely. Weak acids are not completely ionised and a solution of a weak acid has a relatively low concentration of hydrogen ions.

dissolve: to break down a substance in a solution without a resultant reaction.

distillation: the process of separating mixtures by condensing the vapours through cooling.

doping: adding metal atoms to a region of silicon to make it semiconducting.

dye: a coloured substance that will stick to another substance, so that both appear coloured.

electrode: a conductor that forms one terminal of a cell.

electrolysis: an electrical–chemical process that uses an electric current to cause the break up of a compound and the movement of metal ions in a solution. The process happens in many natural situations (as for example in rusting) and is also commonly used in industry for purifying (refining) metals or for plating metal objects with a fine, even metal coating.

electrolyte: a solution that conducts electricity.

electron: a tiny, negatively charged particle that is part of an atom. The flow of electrons through a solid material such as a wire produces an electric current.

electroplating: depositing a thin layer of a metal onto the surface of another substance using electrolysis.

element: a substance that cannot be decomposed into simpler substances by chemical means

emulsion: tiny droplets of one substance dispersed in another. A common oil in water emulsion is milk. The tiny droplets in an emulsion tend to come together, so another stabilising substance is often needed to wrap the particles of grease and oil in a stable coat. Soaps and detergents are such agents. Photographic film is an example of a solid emulsion.

endothermic reaction: a reaction that takes heat from the surroundings. The reaction of carbon monoxide with a metal oxide is an example.

enzyme: organic catalysts in the form of proteins in the body that speed up chemical reactions. Every living cell contains hundreds of enzymes, which ensure that the processes of life continue. Should enzymes be made inoperative, such as through mercury poisoning, then death follows.

ester: organic compounds, formed by the reaction of an alcohol with an acid, which often have a fruity taste.

evaporation: the change of state of a liquid to a gas. Evaporation happens below the boiling point and is used as a method of separating out the materials in a solution.

exothermic reaction: a reaction that gives heat to the surroundings. Many oxidation reactions, for example, give out heat.

explosive: a substance which, when a shock is applied to it, decomposes very rapidly, releasing a very large amount of heat and creating a large volume of gases as a shock wave.

extrusion: forming a shape by pushing it through a die. For example, toothpaste is extruded through the cap (die) of the toothpaste tube.

fallout: radioactive particles that reach the ground from radioactive materials in the atmosphere.

fat: semi-solid energy-rich compounds derived from plants or animals and which are made of carbon, hydrogen and oxygen. Scientists call these esters.

feldspar: a mineral consisting of sheets of aluminium silicate. This is the mineral from which the clay in soils is made.

fertile: able to provide the nutrients needed for unrestricted plant growth.

filtration: the separation of a liquid from a solid using a membrane with small holes.

fission: the breakdown of the structure of an atom, popularly called "splitting the atom" because the atom is split into approximately two other nuclei. This is different from, for example, the small change that happens when radioactivity is emitted.

fixation of nitrogen: the processes that natural organisms, such as bacteria, use to turn the nitrogen of the air into ammonium compounds.

fixing: making solid and liquid nitrogen-containing compounds from nitrogen gas. The compounds that are formed can be used as fertilisers.

fluid: able to flow; either a liquid or a gas.

fluorescent: a substance that gives out visible light when struck by invisible waves such as ultraviolet rays.

flux: a material used to make it easier for a liquid to flow. A flux dissolves metal oxides and so prevents a metal from oxidising while being heated.

foam: a substance that is sufficiently gelatinous to be able to contain bubbles of gas. The gas bulks up the substance, making it behave as though it were semi-rigid.

fossil fuels: hydrocarbon compounds that have been formed from buried plant and animal remains. High pressures and temperatures lasting over millions of years are required. The fossil fuels are coal, oil and natural gas.

fraction: a group of similar components of a mixture. In the petroleum industry the light fractions of crude oil are those with the smallest molecules, while the medium and heavy fractions have larger molecules.

free radical: a very reactive atom or group with a "spare" electron.

freezing point: the temperature at which a substance changes from a liquid to a solid. It is the same temperature as the melting point.

fuel: a concentrated form of chemical energy. The main sources of fuels (called fossil fuels because they were formed by geological processes) are coal, crude oil and natural gas. Products include methane, propane and gasoline. The fuel for stars and space vehicles is hydrogen.

fuel rods: rods of uranium or other radioactive material used as a fuel in nuclear power stations.

fuming: an unstable liquid that gives off a gas. Very concentrated acid solutions are often fuming solutions.

fungicide: any chemical that is designed to kill fungi and control the spread of fungal spores.

fusion: combining atoms to form a heavier atom.

galvanising: applying a thin zinc coating to protect another metal.

gamma rays: waves of radiation produced as the nucleus of a radioactive element rearranges itself into a tighter cluster of protons and neutrons. Gamma rays carry enough energy to damage living cells.

gangue: the unwanted material in an ore.

gas: a form of matter in which the molecules form no definite shape and are free to move about to fill any vessel they are put in.

gelatinous: a term meaning made with water. Because a gelatinous precipitate is mostly water, it is of a similar density to water and will float or lie suspended in the liquid.

gelling agent: a semi-solid jelly-like substance.

gemstone: a wide range of minerals valued by people, both as crystals (such as emerald) and as decorative stones (such as agate). There is no single chemical formula for a gemstone.

glass: a transparent silicate without any crystal growth. It has a glassy lustre and breaks with a curved fracture. Note that some minerals have all these features and are therefore natural glasses. Household glass is a synthetic silicate.

glucose: the most common of the natural sugars. It occurs as the polymer known as cellulose, the fibre in plants. Starch is also a form of glucose. The breakdown of glucose provides the energy that animals need for life.

granite: an igneous rock with a high proportion of silica (usually over 65%). It has well-developed large crystals. The largest pink, grey or white crystals are feldspar.

Greenhouse Effect: an increase of the global air temperature as a result of heat released from burning fossil fuels being absorbed by carbon dioxide in the atmosphere.

gypsum: the name for calcium sulphate. It is commonly found as Plaster of Paris and wallboards.

half-life: the time it takes for the radiation coming from a sample of a radioactive element to decrease by half.

halide: a salt of one of the halogens (fluorine, chlorine, bromine and iodine).

halite: the mineral made of sodium chloride.

halogen: one of a group of elements including chlorine, bromine, iodine and fluorine.

heat-producing: see exothermic reaction.

high explosive: a form of explosive that will only work when it receives a shock from another explosive. High explosives are much more powerful than ordinary explosives. Gunpowder is not a high explosive.

hydrate: a solid compound in crystalline form that contains molecular water. Hydrates commonly form when a solution of a soluble salt is evaporated. The water that forms part of a hydrate crystal is known as the "water of crystallization". It can usually be removed by heating, leaving an anhydrous salt.

hydration: the absorption of water by a substance. Hydrated materials are not "wet" but remain firm, apparently dry, solids. In some cases, hydration makes the substance change colour, in many other cases there is no colour change, simply a change in volume.

hydrocarbon: a compound in which only hydrogen and carbon atoms are present. Most fuels are hydrocarbons, as is the simple plastic polyethene (known as polythene).

hydrogen bond: a type of attractive force that holds one molecule to another. It is one of the weaker forms of intermolecular attractive force.

hydrothermal: a process in which hot water is involved. It is usually used in the context of rock formation because hot water and other fluids sent outwards from liquid magmas are important carriers of metals and the minerals that form gemstones.

igneous rock: a rock that has solidified from molten rock, either volcanic lava on the Earth's surface or magma deep underground. In either case the rock develops a network of interlocking crystals.

incendiary: a substance designed to cause burning.

indicator: a substance or mixture of substances that change colour with acidity or alkalinity.

inert: nonreactive.

infra-red radiation: a form of light radiation where the wavelength of the waves is slightly longer than visible light. Most heat radiation is in the infra-red band.

insoluble: a substance that will not dissolve.

ion: an atom, or group of atoms, that has gained or lost one or more electrons and so developed an electrical charge. Ions behave differently from electrically neutral atoms and molecules. They can move in an electric field, and they can also bind strongly to solvent molecules such as water. Positively charged ions are called cations; negatively charged ions are called anions. Ions carry electrical current through solutions.

ionic bond: the form of bonding that occurs between two ions when the ions have opposite charges. Sodium cations bond with chloride anions to form common salt (NaCl) when a salty solution is evaporated. Ionic bonds are strong bonds except in the presence of a solvent.

ionise: to break up neutral molecules into oppositely charged ions or to convert atoms into ions by the loss of electrons.

ionisation: a process that creates ions.

irrigation: the application of water to fields to help plants grow during times when natural rainfall is sparse.

isotope: atoms that have the same number of protons in their nucleus, but which have different masses; for example, carbon-12 and carbon-14.

latent heat: the amount of heat that is absorbed or released during the process of changing state between gas, liquid or solid. For example, heat is absorbed when a substance melts and it is released again when the substance solidifies.

latex: (the Latin word for "liquid") a suspension of small polymer particles in water. The rubber that flows from a rubber tree is a natural latex. Some synthetic polymers are made as latexes, allowing polymerisation to take place in water.

lava: the material that flows from a volcano.

limestone: a form of calcium carbonate rock that is often formed of lime mud. Most limestones are light grey and have abundant fossils.

liquid: a form of matter that has a fixed volume but no fixed shape.

lode: a deposit in which a number of veins of a metal found close together.

lustre: the shininess of a substance.

magma: the molten rock that forms a balloon-shaped chamber in the rock below a volcano. It is fed by rock moving upwards from below the crust.

marble: a form of limestone that has been "baked" while deep inside mountains. This has caused the limestone to melt and reform into small interlocking crystals, making marble harder than limestone.

mass: the amount of matter in an object. In everyday use, the word weight is often used to mean mass.

melting point: the temperature at which a substance changes state from a solid to a liquid. It is the same as freezing point.

membrane: a thin flexible sheet. A semipermeable membrane has microscopic holes of a size that will selectively allow some ions and molecules to pass through but hold others back. It thus acts as a kind of sieve.

meniscus: the curved surface of a liquid that forms when it rises in a small bore, or capillary tube. The meniscus is convex (bulges upwards) for mercury and is concave (sags downwards) for water.

metal: a substance with a lustre, the ability to conduct heat and electricity and which is not brittle.

metallic bonding: a kind of bonding in which atoms reside in a "sea" of mobile electrons. This type of bonding allows metals to be good conductors and means that they are not brittle

metamorphic rock: formed either from igneous or sedimentary rocks, by heat and or pressure. Metamorphic rocks form deep inside mountains during periods of mountain building. They result from the remelting of rocks during which process crystals are able to grow. Metamorphic rocks often show signs of banding and partial melting.

micronutrient: an element that the body requires in small amounts. Another term is trace element.

mineral: a solid substance made of just one element or chemical compound. Calcite is a mineral because it consists only of calcium carbonate, halite is a mineral because it contains only sodium chloride, quartz is a mineral because it consists of only silicon dioxide.

mineral acid: an acid that does not contain carbon and that attacks minerals. Hydrochloric, sulphuric and nitric acids are the main mineral acids.

mineral-laden: a solution close to saturation.

mixture: a material that can be separated out into two or more substances using physical means.

molecule: a group of two or more atoms held together by chemical bonds.

monoclinic system: a grouping of crystals that look like double-ended chisel blades.

monomer: a building block of a larger chain molecule ("mono" means one, "mer" means part).

mordant: any chemical that allows dyes to stick to other substances.

native metal: a pure form of a metal, not combined as a compound. Native metal is more common in poorly reactive elements than in those that are very reactive.

neutralisation: the reaction of acids and bases to produce a salt and water. The reaction causes hydrogen from the acid and hydroxide from the base to be changed to water. For

example, hydrochloric acid reacts with sodium hydroxide to form common salt and water. The term is more generally used for any reaction where the pH changes towards 7.0, which is the pH of a neutral solution.

neutron: a particle inside the nucleus of an atom that is neutral and has no charge.

noncombustible: a substance that will not burn.

noble metal: silver, gold, platinum, and mercury. These are the least reactive metals.

nuclear energy: the heat energy produced as part of the changes that take place in the core, or nucleus, of an element's atoms.

nuclear reactions: reactions that occur in the core, or nucleus of an atom.

nutrients: soluble ions that are essential to life.

octane: one of the substances contained in fuel.

ore: a rock containing enough of a useful substance to make mining it worthwhile.

organic acid: an acid containing carbon and hydrogen.

organic substance: a substance that contains carbon.

osmosis: a process where molecules of a liquid solvent move through a membrane (filter) from a region of low concentration to a region of high concentration of solute.

oxidation: a reaction in which the oxidising agent removes electrons. (Note that oxidising agents do not have to contain oxygen.)

oxide: a compound that includes oxygen and one other element.

oxidise: the process of gaining oxygen. This can be part of a controlled chemical reaction, or it can be the result of exposing a substance to the air, where oxidation (a form of corrosion) will occur slowly, perhaps over months or years.

oxidising agent: a substance that removes electrons from another substance (and therefore is itself reduced).

ozone: a form of oxygen whose molecules contain three atoms of oxygen. Ozone is regarded as a beneficial gas when high in the atmosphere because it blocks ultraviolet rays. It is a harmful gas when breathed in, so low level ozone, which is produced as part of city smog, is regarded as a form of pollution. The ozone layer is the uppermost part of the stratosphere.

pan: the name given to a shallow pond of liquid. Pans are mainly used for separating solutions by evaporation.

patina: a surface coating that develops on metals and protects them from further corrosion.

percolate: to move slowly through the pores of a rock.

period: a row in the Periodic Table.

Periodic Table: a chart organising elements by atomic number and chemical properties into groups and periods.

pesticide: any chemical that is designed to control pests (unwanted organisms) that are harmful to plants or animals.

petroleum: a natural mixture of a range of gases, liquids and solids derived from the decomposed remains of plants and animals.

pH: a measure of the hydrogen ion concentration in a liquid. Neutral is pH 7.0; numbers greater than this are alkaline, smaller numbers are acidic.

phosphor: any material that glows when energized by ultraviolet or electron beams such as in fluorescent tubes and cathode ray tubes. Phosphors, such as phosphorus, emit light after the source of excitation is cut off. This is why they glow in the dark. By contrast, fluorescors, such as fluorite, emit light only while they are being excited by ultraviolet light or an electron beam.

photon: a parcel of light energy.

photosynthesis: the process by which plants use the energy of the Sun to make the compounds they need for life. In photosynthesis, six molecules of carbon dioxide from the air combine with six molecules of water, forming one molecule of glucose (sugar) and releasing six molecules of oxygen back into the atmosphere.

pigment: any solid material used to give a liquid a colour.

placer deposit: a kind of ore body made of a sediment that contains fragments of gold ore eroded from a mother lode and transported by rivers and/or ocean currents.

plastic (material): a carbon-based material consisting of long chains (polymers) of simple molecules. The word plastic is commonly restricted to synthetic polymers.

plastic (property): a material is plastic if it can be made to change shape easily. Plastic materials will remain in the new shape. (Compare with elastic, a property where a material goes back to its original shape.)

plating: adding a thin coat of one material to another to make it resistant to corrosion.

playa: a dried-up lake bed that is covered with salt deposits. From the Spanish word for beach.

poison gas: a form of gas that is used intentionally to produce widespread injury and death. (Many gases are poisonous, which is why many chemical reactions are performed in laboratory fume chambers, but they are a byproduct of a reaction and not intended to cause harm.)

polymer: a compound that is made of long chains by combining molecules (called monomers) as repeating units. ("Poly" means many, "mer" means part).

polymerisation: a chemical reaction in which large numbers of similar molecules arrange themselves into large molecules, usually long chains. This process usually happens when there is a suitable catalyst present. For example, ethene reacts to form polythene in the presence of certain catalysts.

porous: a material containing many small holes or cracks. Quite often the pores are connected, and liquids, such as water or oil, can move through them.

precious metal: silver, gold, platinum, iridium, and palladium. Each is prized for its rarity. This category is the equivalent of precious stones, or gemstones, for minerals.

precipitate: tiny solid particles formed as a result of a chemical reaction between two liquids or gases.

preservative: a substance that prevents the natural organic decay processes from occurring. Many substances can be used safely for this purpose, including sulphites and nitrogen gas.

product: a substance produced by a chemical reaction.

protein: molecules that help to build tissue and bone and therefore make new body cells. Proteins contain amino acids.

proton: a positively charged particle in the nucleus of an atom that balances out the charge of the surrounding electrons

pyrite: "mineral of fire". This name comes from the fact that pyrite (iron sulphide) will give off sparks if struck with a stone.

pyrometallurgy: refining a metal from its ore using heat. A blast furnace or smelter is the main equipment used.

radiation: the exchange of energy with the surroundings through the transmission of waves or particles of energy. Radiation is a form of energy transfer that can happen through space; no intervening medium is required (as would be the case for conduction and convection).

radioactive: a material that emits radiation or particles from the nucleus of its atoms.

radioactive decay: a change in a radioactive element due to loss of mass through radiation. For example uranium decays (changes) to lead.

radioisotope: a shortened version of the phrase radioactive isotope.

radiotracer: a radioactive isotope that is added to a stable, nonradioactive material in order to trace how it moves and its concentration.

reaction: the recombination of two substances using parts of each substance to produce new substances.

reactivity: the tendency of a substance to react with other substances. The term is most widely used in comparing the reactivity of metals. Metals are arranged in a reactivity series.

reagent: a starting material for a reaction.

recycling: the reuse of a material to save the time and energy required to extract new material from the Earth and to conserve non-renewable resources.

redox reaction: a reaction that involves reduction and oxidation.

reducing agent: a substance that gives electrons to another substance. Carbon monoxide is a reducing agent when passed over copper oxide, turning it to copper and producing carbon dioxide gas. Similarly, iron oxide is reduced to iron in a blast furnace. Sulphur dioxide is a reducing agent, used for bleaching bread.

reduction: the removal of oxygen from a substance. See also: oxidation.

refining: separating a mixture into the simpler substances of which it is made. In the case of a rock, it means the extraction of the metal that is mixed up in the rock. In the case of oil it means separating out the fractions of which it is made.

refractive index: the property of a transparent material that controls the angle at which total internal reflection will occur. The greater the refractive index, the more reflective the material will be.

resin: natural or synthetic polymers that can be moulded into solid objects or spun into thread.

rust: the corrosion of iron and steel.

saline: a solution in which most of the dissolved matter is sodium chloride (common salt).

salinisation: the concentration of salts, especially sodium chloride, in the upper layers of a soil due to poor methods of irrigation.

salts: compounds, often involving a metal, that are the reaction products of acids and bases. (Note "salt" is also the common word for sodium chloride, common salt or table salt.)

saponification: the term for a reaction between a fat and a base that produces a soap.

saturated: a state where a liquid can hold no more of a substance. If any more of the substance is added, it will not dissolve.

saturated solution: a solution that holds the maximum possible amount of dissolved material. The amount of material in solution varies with the temperature; cold solutions can hold less dissolved solid material than hot solutions. Gases are more soluble in cold liquids than hot liquids.

sediment: material that settles out at the bottom of a liquid when it is still.

semiconductor: a material of intermediate conductivity. Semiconductor devices often use silicon when they are made as part of diodes, transistors or integrated circuits.

semipermeable membrane: a thin (membrane) of material that acts as a fine sieve, allowing small molecules to pass, but holding large molecules back.

silicate: a compound containing silicon and oxygen (known as silica).

sintering: a process that happens at moderately high temperatures in some compounds. Grains begin to fuse together even through they do not melt. The most widespread example of sintering happens during the firing of clays to make ceramics.

slag: a mixture of substances that are waste products of a furnace. Most slags are composed mainly of silicates.

smelting: roasting a substance in order to extract the metal contained in it.

smog: a mixture of smoke and fog. The term is used to describe city fogs in which there is a large proportion of particulate matter (tiny pieces of carbon from exhausts) and also a high concentration of sulphur and nitrogen gases and probably ozone.

soldering: joining together two pieces of metal using solder, an alloy with a low melting point.

solid: a form of matter where a substance has a definite shape.

soluble: a substance that will readily dissolve in a solvent.

solute: the substance that dissolves in a solution (e.g. sodium chloride in salt water).

solution: a mixture of a liquid and at least one other substance (e.g. salt water). Mixtures can be separated out by physical means, for example by evaporation and cooling.

solvent: the main substance in a solution (e.g. water in salt water).

spontaneous combustion: the effect of a very reactive material beginning to oxidise very quickly and bursting into flame.

stable: able to exist without changing into another substance.

stratosphere: the part of the Earth's atmosphere that lies immediately above the region in which clouds form. It occurs between 12 and 50 km above the Earth's surface.

strong acid: an acid that has completely dissociated (ionised) in water. Mineral acids are strong acids.

sublimation: the change of a substance from solid to gas, or vica versa, without going through a liquid phase.

substance: a type of material, including mixtures.

sulphate: a compound that includes sulphur and oxygen, for example, calcium sulphate or gypsum.

sulphide: a sulphur compound that contains no oxygen.

sulphite: a sulphur compound that contains less oxygen than a sulphate.

surface tension: the force that operates on the surface of a liquid, which makes it act as though it were covered with an invisible elastic film.

suspension: tiny particles suspended in a liquid.

synthetic: does not occur naturally, but has to be manufactured.

tarnish: a coating that develops as a result of the reaction between a metal and substances in the air. The most common form of tarnishing is a very thin transparent oxide coating.

thermonuclear reactions: reactions that occur within atoms due to fusion, releasing an immensely concentrated amount of energy.

thermoplastic: a plastic that will soften, can repeatedly be moulded it into shape on heating and will set into the moulded shape as it cools.

thermoset: a plastic that will set into a moulded shape as it cools, but which cannot be made soft by reheating.

titration: a process of dripping one liquid into another in order to find out the amount needed to cause a neutral solution. An indicator is used to signal change.

toxic: poisonous enough to cause death.

translucent: almost transparent.

transmutation: the change of one element into another.

vapour: the gaseous form of a substance that is normally a liquid. For example, water vapour is the gaseous form of liquid water.

vein: a mineral deposit different from, and usually cutting across, the surrounding rocks. Most mineral and metal-bearing veins are deposits filling fractures. The veins were filled by hot, mineral-rich waters rising upwards from liquid volcanic magma. They are important sources of many metals, such as silver and gold, and also minerals such as gemstones. Veins are usually narrow, and were best suited to hand-mining. They are less exploited in the modern machine age.

viscous: slow moving, syrupy. A liquid that has a low viscosity is said to be mobile.

vitreous: glass-like.

volatile: readily forms a gas.

vulcanisation: forming cross-links between polymer chains to increase the strength of the whole polymer. Rubbers are vulcanised using sulphur when making tyres and other strong materials.

weak acid: an acid that has only partly dissociated (ionised) in water. Most organic acids are weak acids.

weather: a term used by Earth scientists and derived from "weathering", meaning to react with water and gases of the environment.

weathering: the slow natural processes that break down rocks and reduce them to small fragments either by mechanical or chemical means.

welding: fusing two pieces of metal together using heat.

X-rays: a form of very short wave radiation.

Index